Dark Matters

Dark Matters

Unifying Matter, Dark Matter, Dark Energy, *and the* Universal Grid

Dr. Percy Seymour

New Page Books
A Division of The Career Press, Inc.
Franklin Lakes, NJ

DARK MATTERS
EDITED AND TYPESET BY KARA REYNOLDS
Cover design by Lu Rossman/Digi Dog Design NY
Printed in the U.S.A. by Book-mart Press

To order this title, please call toll-free 1-800-CAREER-1 (NJ and Canada: 201-848-0310) to order using VISA or MasterCard, or for further information on books from Career Press.

The Career Press, Inc., 3 Tice Road, PO Box 687,
Franklin Lakes, NJ 07417
www.careerpress.com
www.newpagebooks.com

Library of Congress Cataloging-in-Publication Data

Seymour, Percy.
Dark matters : unifying matter, dark matter, dark energy, and the universal grid / by Dr. Percy Seymour.
p. cm.
Includes bibliographical references and index.
ISBN-13: 978-1-60163-006-3
1. Dark matter (Astronomy) 2. Dark energy (Astronomy) I. Title.

QB791.3.S49 2008
523.1--dc22

2007046528

To my wife, Dianna,
and my son, Bruce

Acknowledgments

First I would like to thank my wife, Dianna, for putting up with the birth and development of my theories all the time we have been together, and my son, for putting up with these theories for much of his early years. I would also like to thank my friend Andrew Wallace for modifying and improving some of my diagrams. My agent Lisa Hagan has been a great support to me in encouraging my writing efforts and finding me a publisher for the present book. I was thrilled when Michael Pye of New Page Books/Careers Press offered me a contract, and I must thank him for his interest in my work. I would also like to thank the following members of the editorial staff of Career Press for their patience and assistance during the editorial process: Kirsten Dalley, Kara Reynolds, and Kristen Parkes.

Contents

Introduction

The Dark Side of the Cosmos

There is a dark side to the universe. Evidence for the darkness of the cosmos has been growing at an increasing rate throughout the past few decades, but the rate of growth has increased considerably in the last two to three years. Unfortunately, the accumulation of data concerning these unseen components of the universe has not been matched by an understanding of what is giving rise to these constituents of the world. There are two components involved: One is called *dark matter* and the other is called *dark energy*. The evidence for dark matter comes from the behavior of stars and other matter in those vast collections of stars called galaxies, and from the behavior of matter in large collections of galaxies called clusters. The evidence for dark energy comes from the observations that the most distant galaxies seem to be accelerating away from each other at an increasing rate. This book will explore a theory that could explain both components, but which also entails a radical revision of our current theories of matter. We will start with a look at some recent comments on these puzzling aspects of modern physics, astronomy, and cosmology.

COMMENTS ON DARK MATTER AND DARK ENERGY

In the June 2006 issue of *Physics World* (a magazine circulated to members and fellows of the Institute of Physics in the UK) there was an article called "Gravity's Dark Side." It drew attention to the fact that, since 1933, astronomical observations have been leading us to the conclusion that there is an "invisible" component of matter now known as *dark matter.* The article quoted HongSheng Zhao of St. Andrews University as saying, "Astronomers have no idea what dark matter is.... It is whatever is needed to explain the data, rather than a fundamental prediction of particle physics as it was originally." Some astronomers and physicists are seeking to solve the problem by modifying Einstein's general relativity, which is currently our best theory of gravitation.

In another part of the same issue of *Physics World* there is an item called "Physicists Eye Dark Energy Matters." Dark energy has been evoked to explain the observed fact that the expansion of the universe is accelerating. In the United States, a Dark Energy Task Force has been set up to investigate the matter. Rocky Kolh, chair of the task force, said, "Dark energy ranks among the most compelling of all the outstanding problems in physics.... These circumstances demand an ambitious and coordinated program to determine its properties as well as possible."

In the August 2006 issue of *A&G* ("News and Reviews" in *Astronomy and Geophysics*, a magazine circulated to fellows and members of the Royal Astronomical Society in England), there was an article by Roberto Trotto and Richard Bower called "Surveying the Darker Side," starting with the words: "One of the most fundamental problems of contemporary physics is the nature of the dark sector of the universe: dark matter and dark energy."

The article had these words near the end: "...the observational study of dark energy is a crucial area of cosmological research...one of the most challenging problems in...physics...likely to spark a new understanding of fundamental physics."

The items quoted here are just a few of several that appeared in newspapers, magazines, radio shows, and television programs in 2006. Dark matter was also reviewed in the *Astronomy Now 2007 Yearbook.* Here, in an article called "Dark Matters," the reviewer draws attention to some work by Professor Gerry Gilmore (from the Institute of Astronomy at Cambridge University), who, with his team, mapped the distribution of dark matter across 12 dwarf galaxies, and concluded that dark matter had a temperature of 10,000 degrees Celsius. However, Professor Gilmore added: "This is not a real temperature...." The reviewer clarified this comment: "What he means is that if dark matter 'particles' have the same mass as a proton, this is what their temperature would be."

"In reality we don't even know what mass they are," says Gilmore. "They may not even be particles." Further along the reviewer said, "So there are no firm answers at present, just hints, ideas, and possibilities."

The January 2007 issue of *Physics World* carried a report of some recent work on dark energy. The Hubble Space Telescope was used by Adam Reiss and colleagues of John Hopkins University to shed light on the nature of this unseen component of energy. Their results revealed that "dark energy" was in existence about nine billion years ago: "The data suggest that the effect of dark energy was rather weak until about five to six billion years ago, when it defeated gravity in a 'cosmic tug of war,' causing the rate of expansion to increase."

Before we can begin to appreciate these problems, we need to set the scene by highlighting, very briefly, some aspects of modern physics.

The Building Blocks of the Universe

The idea that all matter was made of small particles started in the ancient world. According to Leucippus, who flourished around 430 BC, and his pupil, Democritus (c. 460–371 BC), matter did not form a continuum, but consisted of eternal, invariant, impenetrably hard, homogeneous parts called *atoms*, which moved in a void. This view was not generally accepted, and among those who opposed it were Plato (427–347 BC) and Aristotle (384–322 BC).

These early ideas on atoms were developed into a full scientific theory that was to form the basis of the modern approach to chemistry by the Manchester chemist John Dalton (1766–1822). Dalton proposed that chemical elements, which could not be broken down into anything more basic by chemical means, consisted of atoms. However, the atoms of one element could be fused with the atoms of another element to form a chemical compound. At this stage there was no evidence that atoms could be broken down into smaller particles, so, from Dalton's point of view, each element had its own type of indivisible atom.

The first evidence that there were particles smaller than atoms came from the work of J.J. Thomson (1856–1940), who discovered the electron at Cambridge University in 1897. Thomson proposed a model for atomic structure, but there were serious problems with it. The first steps toward a better model of the atom, based on actual experiments, were made by Ernest Rutherford (1871–1937), working at Manchester University, from 1908 to 1919. Rutherford set two of his researchers, Hans Geiger (1882–1945) and Ernest Marsden (1889–1970), the task of firing certain emissions from a radioactive substance at a thin foil of gold, and then getting them to measure the angles through which the particles were deflected. Some of these particles seemed to bounce right back from the foil, and this led Rutherford to conclude that the atom was rather like a miniature solar system, with a very small and dense central nucleus with electrons orbiting around it, like the planets orbiting the sun. By the time I entered Manchester University, in 1962, to study physics, Rutherford's research laboratory had become an undergraduate laboratory, and sometimes I carried out simple physics experiments at the bench, now marked by a plaque, at which Geiger and Marsden had carried out this important experiment. This was where the atomic age really started! It always gave me a great thrill to think that I sometimes worked where one of the greatest experiments of the 20th century was carried out.

Unfortunately, Rutherford's model of the atom was problematic from the point of view of physics as it was understood at that time. The laws of physics that existed before 1900 were known as the principles of classical physics. According to these laws, the electron would radiate energy as it circled the central nucleus, and this loss of energy would cause it to spiral into the nucleus. This problem was solved by Niels Bohr, who applied the rules of quantum physics (which were just beginning to appear) to Rutherford's model of the atom. In the case of the hydrogen atom he was able to show that its single electron could, according to quantum principle, only orbit at certain fixed distances from the nucleus. When the electron was in one of these "fixed orbits" it would not radiate, but when it jumped from one orbit to the next, it would either radiate some energy or absorb some energy.

Rutherford, in 1919, had identified the nucleus of the hydrogen atom as a new particle, which he called the proton. This particle had a positive electric charge, of the same value as the charge on the electron, but with an opposite sign. Its mass was almost 2,000 times more than that of the electron. In 1932 James Chadwich (1891–1974) discovered a subatomic particle called a *neutron*. This particle had no charge, but its mass was virtually the same as that of the proton. The stage was now set to understand how the chemical elements were built up: All the 98 naturally occurring chemical elements were made up of just three particles, mixed together in different proportions. The different masses of the atoms of each element came, mainly, from the fact that they had different numbers of protons and neutrons in their nuclei. Neutral atoms had the same number of electrons orbiting the central nucleus as there were protons in the nucleus. The varying chemical properties of atoms arose from the number of electrons orbiting the nucleus, and how they were arranged into different "orbital shells."

The Cement and Mortar of the Universe

In the last section we met the three basic building blocks of most ordinary atoms of the elements. But what holds them together? And how do they combine to give rise to the diversity of chemical compounds, minerals, life, planets, stars, galaxies, and the whole universe? In this section I will sketch the main forces that are responsible for the building process.

Throughout the last 3,000 years scientists have discovered five different types of forces, some of which can be attractive *and* repulsive, and others that are normally only attractive. Modern physicists have classified these forces as follows:

- The electromagnetic force.
- Gravitation.
- The strong nuclear force.
- The weak nuclear force.

Let us explore each of these in just a bit more detail.

The Electromagnetic Force

Most of us have some experience with static electricity. For example, rubbing two pieces of amber together will allow one to pick up small pieces of paper with either of the pieces of amber—this is static electricity, which was first discovered by the ancient Greek scientist Thales of Miletus (624–547 BC).

The Greeks were also aware that there existed an ore, called lodestone, that could attract iron. In China, starting about 700 years ago, people practiced geomancy, which taught that in order to have health and happiness in this world, and to have a peaceful one in the next, one had to take full advantage of the winds, waters, fields, and forces of the earth. This meant that one had to sleep and do one's common tasks in the "right direction," and when one died one had to be buried in an "appropriate direction." To carry out these procedures it was necessary to have a device that could be used to find direction, so the practitioners of geomancy invented the geomantic compass, which came to be the forerunner of the mariner's compass.

The next big leap forward in the study of magnetism came from the British scientist William Gilbert (1540–1603), who was not only physician to Queen Elizabeth the First, but who was also one of the founding fathers of experimental science. In 1600, Gilbert published a book called *De Magnete*, in which he discussed the basic attributes of magnetism, all based on his own experimental investigations. Using data obtained by mariners who had gone to other parts of the world, he came to the conclusion that Earth behaved as if it had a large magnet on the inside, and that the poles of this magnet were close to the geographic poles of the planet. Gilbert also used *electrum* (the Latin word for "amber"), to coin the term *electrica* to describe amber's property of attracting lighter objects.

At this stage electricity and magnetism were still seen as separate entities. In 1820, the French scientist A.M. Ampere (1775–1836), verified that a magnetic force is generated when an electric current passes through a coil of conducting wire. He also proposed that the magnetism of substances, such as iron, nickel, and cobalt, arose from electric currents circulating on a molecular scale. Michael Faraday (1791–1854) discovered that if an instrument, called an ammeter, was connected to the ends of a coil of wire, and a bar magnet moved in and out of the coil, then an electric current would pass through the completed circuit. These experiments showed that electricity and magnetism were linked together, and that they were not separate entities. The Scottish physicist James Clerk Maxwell (1831–1879) used the experimental work of Faraday to work out the mathematical theory electromagnetism, which combined electricity with magnetism in an intimate way, and showed that light waves consisted of varying electric and magnetic fields that propagated through empty space at a fixed speed, now known as the speed of light. In 1905 Albert Einstein (1879–1955) made the speed of light one of the cornerstones of his special theory of relativity, and showed that nothing could travel faster than the speed of light.

Atoms are held together by the force of static electricity that exists between the orbiting electrons, which have a negative charge, and the central nucleus, which has a positive charge. Charles Augustin de Coulomb (1736–1806) discovered the law that now bears his name: Coulomb's law. This law tells us that like charges repel each other, and unlike charges attract each other. The force of attraction (or the force of repulsion) between two particles also gets weaker as the distance between them is increased. If the distance is doubled then the force of attraction (or repulsion) is four times weaker. This is called Coulomb's inverse square law.

The electrical interactions between the electrons of one chemical element and those of another chemical element are responsible for the combining of elements to form chemical compounds. The forces that combine all solid materials together are also associated with the behavior of the electrons. Thus, on the everyday level of existence, the solid feel of matter is basically electrical in origin—or, more generally, electromagnetic in origin.

Gravitation

Although gravitation is the weakest of all the forces, it is the dominant force on the scale of asteroids, planets, stars, and galaxies. Newton's law of gravitation tells us that every particle in the universe attracts every other particle. It also tells us that the force of attraction, similar to that between oppositely charged particles, obeys the inverse square law, and thus if the distance between two particles is doubled, then the force of attraction is four times weaker. When we are dealing with the millions of millions of particles in bodies of astronomical dimensions, then gravitation is the dominant force. Whereas like electrical charges repel and unlike electrical charges attract (and like magnetic poles repel and unlike ones attract), on the scale of planets, stars, and galaxies, gravitation is always attractive.

Weight and mass are different. The mass of a body is a measure of how much material there is in it, and this is the same no matter where in the universe the body is situated. The weight of a body on the surface of the Earth is due to the gravitational tug of all the particles of which the Earth is composed, on the particles of which the body is composed. This means it depends on the total mass of the Earth, and for a spherical body it acts as if all the mass of the Earth was concentrated at the center.

The planets orbit the sun under the control of gravitation, our moon orbits the Earth, and the satellites of those planets that have them are all under the control of the universal force of gravitation. According to Isaac Newton's first law of motion, all bodies will remain at rest, or they will move in straight lines (if they are already moving), unless they are acted on by a force that will change the state of rest or motion in a straight line. Thus, our moon has the natural tendency to want to move off into space in a straight line, which is at 90 degrees to the line joining its center to that of Earth. What stops it doing that is the gravitational tug between the moon and the Earth. The speed with which the moon goes around the Earth depends on its distance from Earth and the total mass of Earth. This means that the moon's motion around the Earth can be used to find the mass of the Earth. This method can also be used to find the masses of planets with satellites.

Throughout the years astronomers have found a large number of binary systems, which are two stars orbiting around each other. In the simplest case, in which a less massive star orbits a more massive one, astronomers can measure changes in the light from the smaller one to determine the speed with which it is orbiting the larger one, and combining this with measurements of how far apart they are, the mass of the larger companion can be calculated. The same method can be used to find the masses of galaxies. Once again, by measuring changes in the light from the outermost stars of a galaxy, astronomers can find the speeds of these stars, and knowing how far they are from the center of the galaxy, it is possible to work out the mass of the galaxy. Some of the first evidence for dark matter came from such studies on galaxies. Astronomers could estimate the total mass of a galaxy by studying the numbers of different types of stars in the galaxy, and knowing the average masses of each type of star, they could add the whole lot together to calculate the mass of the galaxy. This turned out to be lower than the mass measured by studying the light form the stars in the outermost parts of the galaxy. (All of this will be discussed in more detail in Chapter 5.)

Evidence for dark energy comes from a different source. In 1916 Albert Einstein formulated his general theory of relativity, which was in essence a new theory of gravitation. This theory opened up the possibility of discussing the large-scale structure of the whole universe in rigorous mathematical terms. The general theory was much more powerful than Newton's theory of gravitation, but the main mathematical terms of the theory still required gravitation to be, solely, an attractive force. At this stage galaxies seemed to be keeping their distances with respect to each other, which was not what one would have expected if there were attractive gravitational forces between them. Einstein noticed that his equation did allow for the possibility of a repulsive term, which would only operate when the galaxies were sufficiently far apart. This came to be known as the cosmological repulsive term. By including this term when discussing the large-scale evolution of the universe, he could explain why the galaxies were static with respect to each other. In 1929 Edwin Hubble (1889–1953), by analyzing the light from distant galaxies, found that the most distant galaxies were moving away from each other at speeds that were directly related to their distances apart. Thus, those furthest from us were moving at the highest speeds. This provided evidence for the theory that the universe started with a big bang. The most distant galaxies would be traveling at the highest speeds because they were given the "biggest push" by the initial explosion. The law relating the speeds of galaxies to their distances apart is called Hubble's law, and it made it unnecessary to invoke Einstein's repulsive cosmological term. However, observations made in the last few years have revived interest in this term. It is these observations that provide evidence for dark energy.

Astronomers have been using a special type of star, known as type 1a supernovae, to measure the distances to some of the most distant objects in the universe. These stars are being used as standard candles: objects for which the actual brightness is known, so it is possible to relate the observed brightness to the distance. When this method was applied to the most distant objects it was found that these objects were receding from us at a rate greater than that which one would expect according to Hubble's law. This implied that there was an entity accelerating them, which was acting in opposition to the gravitational attraction between them. It is this entity that has been given the name of *dark energy*.

The Strong Nuclear Force

This is the force that holds the particles of a nucleus together. The hydrogen atom consists of one proton in the nucleus, and one electron orbiting the nucleus, so there is no need for the strong nuclear force to play a role. However, the helium atom has two protons and two neutrons in its nucleus, with two electrons orbiting it, so some strong, additional force is needed to hold it together. Because the two protons both have positive electrical charges they will tend to push each other apart. The force that counteracts this is called the *strong nuclear force.* It is, however, a short-range force, and only comes into play when the protons and neutrons are very close together. It is almost as if the strong nuclear force is like a strong superglue, which keeps the particles of the nucleus together, when they are close enough for it to come into effect.

The Weak Nuclear Force

This force comes into play to convert a neutron into a proton, when the neutron, which is a short-lived particle when outside the nucleus (lasting only for about a quarter of an hour), decays into a proton, an electron, and an anti-neutrino. In the nucleus it plays the role of converting a proton into a neutron, a positron (a positively charged electron), and a neutrino. Aspects of all these topics will be developed in the rest of the book.

The Genesis of a New Theory of Matter

In 1989 I formulated a new theory of matter. I have always been proud of the fact that I studied physics at Manchester University. The physics, mathematics, and astronomy departments of the university had a character that was very much in keeping with the character of science and technology of the city of Manchester itself. Science was not theoretical science for its own sake; initially the purpose of science was to serve the Industrial Revolution, so it was practical, and focused on useful results.

Manchester was the place where the Industrial Revolution started. It was also the world's first atomic city. It was here that John Dalton (1766–1844), a chemist, turned the Greek idea about atoms into a full scientific theory that was to form a new basis for chemistry. It was here that James Joule (1818–1889) established relationships between electrical current, mechanical energy, and heat. It was here that Ernest Rutherford (1871–1937) supervised experiments that established that the atom consisted of a dense central nucleus surrounded by orbiting electrons. Niels Bohr (1885–1962), the Danish physicist, came to England to work under J.J. Thomson, Cavendish professor at Cambridge University, but the two men had a personality clash, so he went to Manchester to work with Rutherford. It was here that the Rutherford-Bohr model of the atom was first conceived.

It was not only in the physics department that experimental and practical considerations formed the basis of theories. Even the mathematics department concentrated largely on problems in engineering, fluid mechanics, and aerodynamics.

In my undergraduate course in physics I was introduced to the mathematical precision of the theories of relativity and quantum mechanics, including the initial stages of relativistic quantum mechanics. However, for my post-graduate studies in astronomy I worked on the role of magnetic fields in astronomical objects. The contrast between the two areas of research was vast. The problems concerning the interactions between magnetic fields and gas motions in cosmic situations were so complex that the majority of them did not lend themselves to rigorous mathematical solutions. Although applied mathematicians had given substance to the pressures and tensions that Michael Faraday had conceived to be associated with magnetic lines of force, most of the problems encountered when working on magnetic fields in a cosmic context could only be dealt with by approximate methods. This was not just due to academic or technical difficulties; it was inherent in the very nature of the problems.

In physics, such problems were called *nonlinear*. The general area of astronomical research that involves the linking-up of the mathematical equations of electromagnetism with those of gas dynamics or hydrodynamics is called *cosmic electrodynamics*. This area of research tells us that the flow of the fluids alters the structure of the magnetic fields, and this in turn alters the flow of the fluids, so we can only proceed by successive approximation.

Although most of my research and teaching since I left Manchester in 1972 has been concerned with astronomy and astrophysics, I still try to keep abreast of developments in other areas of physics. However, I was uneasy about the way subatomic physics was going. I was uncomfortable with the quark theory of matter, and when string theory came along I was even more uncomfortable, especially as it involved spaces of more than four dimensions.

Ever since my first introduction to neutrons, protons, and electrons, I was struck by the fact that although the proton was almost 2,000 times more massive than the electron, the electric charge on the two were exactly equal in magnitude, but opposite in sign! As a post-graduate, I was attending lectures on cosmic electrodynamics in the department of mathematics at Manchester University (given by Professor Leon Mestel, who is a world authority on magnetic stars), and lectures on magnetic fields in astronomy (given by the late Professor Franz Kahn, a world authority on the interstellar medium). I had an idea about how the neutron might decay to give a proton, an electron, and a anti-neutrino. The image I conceived was that this neutron decay was similar, in some respects, to the formation of a loop prominence and a sunspot pair on the Sun. It was several years before I could take this idea further. (In Chapter 7 I will explain this theory in more detail.) The central feature of this theory is the concept of lines of force. In this theory there are no particles, only three kinds of space and electric lines of force. What we normally call *particles* are compacted bundles of lines of force, wound up in the form of balls of yarn. The electrical energy in these balls of yarn give the particles their masses according to Einstein's formula $E = mc^2$. In *ordinary space* these lines will go from a proton to an electron, but in *plasma space* the lines of force go back from the electron to the proton in very thin, braided ropes encased in sheaths of *insulating space*. The sheaths of insulating space are elasticated, so they behave like stretched elastic tubes or spiral springs. Their stretching means that they have energy stored in them. At the atomic and subatomic level, the main consequence of these encased ropes of electric lines of force create (when we are dealing with large numbers of protons and electrons) a criss-crossing web that can push particles around, thus giving rise to the so-called quantum noise of the universe. On the scale of galaxies and clusters of galaxies, the many billions upon billions of these ropes, and the tension in the sheaths surrounding them, give rise to an additional attraction between vast masses of particles. It is this additional attraction that we call *dark matter*. The energy contained in the web itself gives rise to additional mass, which is another component of dark matter. On the cosmological scale the braiding begins to unbraid as the cosmic net expands and the lines of force within the sheaths of insulating space become more and more parallel to each other. The tension, or negative pressure, in the elasticated sheaths will increase as the universe expands, and this, in the context of the general theory of relativity, will give rise to the expansion of the universe. (This will be discussed further in Chapter 7.)

Part I
Magnetic Threads of the Cosmic Tapestry

Chapter One

Michael Faraday and His Lines of Reasoning

Electricity is a dominant feature of our lives in the modern world: Electric motors drive our washing machines, our vacuum cleaners, and the starters of our cars. The electric power to our homes is generated in power stations by very large generators. High-voltage power cables are stepped down to lower voltages by transformers. All these devices are developments of inventions by Michael Faraday. He was one of the greatest experimental physicists of all time. He had about a dozen laws, effects, and inventions named after him. However, one of his greatest contributions to science was a way of looking at fields and forces that completely transformed the theoretical basis of the physical sciences. In this chapter we will look at his life and work, but we will pay particular attention to his concept of lines of force and their associated physical properties.

FARADAY'S LIFE IN BRIEF

Michael Faraday was born on September 22, 1791, in a part of London then called Newington Butts, now known as Elephant and Castle. Faraday's parents had come to London from the northwest of England in the late 1780s. His father was a blacksmith, and both parents belonged to the Sandemanian sect of Christianity. He left school at the age of about 13, but not before having learned the "rudiments of reading, writing and arithmetic."

At the age of 14 Faraday started a seven-year apprenticeship, as a bookbinder, to George Riebau of Blandford Street in London. Faraday was allowed to read the books he bound. He was particularly attracted to scientific articles in the *Encyclopaedia Britannica*, and he was much influenced by the book *Conversations in Chemistry* by Jane Mercet. In 1810 he joined the City Philosophical Society, a group of men and women who met every week to hear, specifically, lectures on scientific subjects, and, more generally, to discuss matters of scientific interest. It was at this society that Faraday was to give his own first scientific lectures.

In 1812, William Dance, one of the customers of George Riebau, gave Faraday four tickets to hear four lectures by Humphry Davy at the Royal Institution in London. During these lectures the young Faraday wrote detailed notes, and back at home he made a careful transcript from the notes and bound them into a book. He sent this book to Davy, together with a letter asking if there were any vacancies at the Royal Institution. Davy invited him for an interview, but told him that there were no vacancies at present. However, a few months later, early in 1813, the chemical assistant was dismissed, following a fight with the instrument maker in the main lecture theater, and Humphrey Davy offered Faraday the position, which he accepted.

Between 1810 and 1820, and for most of the 1820s, Faraday worked under the professor of chemistry, William Thomas Brande, at the Royal Institution. For several months during this period, from October 1813 to April 1815, he went (with Davy, Davy's wife, and Davy's maid) as an assistant and valet on a scientific tour of Europe. They visited France, Italy, Switzerland, and Southern Germany. On his return to England he resumed his post as chemical assistant.

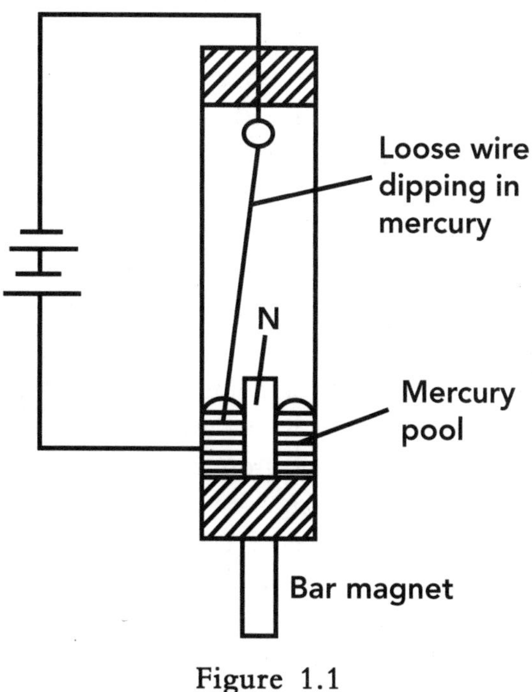

Figure 1.1
Faraday's rotation experiment.

At this time there was a great deal of interest in the discovery by Danish natural philosopher Hans Christian Oersted (1777–1851) that a current-carrying conductor was surrounded by a magnetic field. Faraday also became very interested in this discovery, and in September 1821 he made his own contribution to the interaction between the fields of a current-carrying conductor and one of the poles of a permanent bar magnet. He arranged for one end of a straight conductor to be attached at its upper end to another conductor, in such a way that its other end was free to move about in a bowl of mercury, into which this other end dipped. Protruding into the center of the bowl of mercury was one pole of a bar magnet. When the current was switched on, the moveable straight conductor orbited the bar magnet (see Figure 1.1). With this experiment he had discovered electromagnetic rotation, which is the principle of the electric motor.

For a period of 10 years following this discovery, he was very much engaged with other duties connected with the Royal Institution, and during this time he did not carry out much research, but in August 1831 he made two more discoveries of great importance. Both of these were to do with electromagnetic induction. In this first experiment he showed that if a bar magnet was moved in and out of a coil of wire, then an electric current was induced in the coil. This is the basic principle of the electric generator. In the second experiment he had two separate coils of wire (covered with insulating material, except at their ends) wrapped around a bar of iron. When the current in the one coil was switched on and off, a current was induced in the other coil. This is the principle of the electric transformer.

Faraday, in 1836, was appointed as scientific adviser to Trinity House, the organization that administers the light houses around the shores of England and Wales, and generally looks after marine safety, and continued in this capacity until 1865, two years before he died. From 1830 to 1851 he was also professor of chemistry at the Royal Military Academy at Woolwich in Greater London. These are two examples of how he was asked to apply his scientific abilities to practical ends.

His scientific output went into a slight decline in the early 1940s due to ill health and his appointment as an Elder in the Sandemanian Church. However, in 1843, he became interested in the electrical conductivity of space itself, as distinct from different types of matter. He discussed this subject in 1844, in a lecture on the nature of matter, in which he proposed that atoms should be seen as centers of force where lines of force met, rather than as hard, solid spheres, as had been proposed by chemist John Dalton. Also during this period, he investigated the effect of magnetic fields on the polarization of light, and discovered the Faraday Effect, which we will discuss in more detail later in this chapter. In April 1846 he delivered a lecture he called "Thoughts on Ray-Vibrations," in which he laid the basis for the field theory of electromagnetism. This theory was first taken up by William Thomson (1824–1907), and then by James Clerk Maxwell. It was Maxwell who formulated the full theory of electromagnetism, which unified electricity, magnetism, and light. Subsequently, as new types of radiation were discovered—X-rays, radio waves, infrared, and ultraviolet light—it was realized that they too were part of what is now known as the electromagnetic spectrum.

Queen Victoria granted Faraday a "grace and favour house" (for which no rent is paid) at Hampton Court Palace (originally the Royal Palace of King Henry VIII), in1858, and this is where he died, on August 25, 1867.

Faraday: The Natural Philosopher

Faraday's method of thinking had a great effect on James Clerk Maxwell, the formulator of electromagnetic theory. In his book *Einstein*, Banesh Hoffmann tells us:

> Meanwhile, however, the English experimenter Michael Faraday was making outstanding experimental discoveries in electricity and magnetism. Being largely self-taught and lacking mathematical facility, he could not interpret his results in the manner of Ampere. And this was fortunate, since it led to a revolution in science. Ampere and others had concentrated their attention on the visible hardware—magnets, current-carrying wires, and the like—and on the numbers of centimeters separating the pieces of hardware. In so doing they were following the action-at-a-distance tradition that had developed from the enormous success of the Newtonian system of mechanics and law of gravitation.

Further along Hoffmann says:

> But Faraday regarded the hardware as secondary. For him the important physical events took place in the surrounding space—the field. This, in his mind, he filled with tentacles that by their pulls and thrusts and motions gave rise to the electromagnetic effects observed. Although he could thus interpret his electromagnetic experiments with excellent precision and surprising simplicity, most physicists adept at mathematics thought his concepts mathematically naïve.

George Biddell Airy, who was the Astronomer Royal at Greenwich at that time, was one of the scientists who did not think highly of Faraday's *lines of force* concept. On the other hand, James Clerk Maxwell, the Scottish theoretical physicist, gave Faraday's ideas very serious consideration. He said, in *A Treatise on Electricity and Magnetism*: "As I proceeded with the study of Faraday, I perceived that his method of conceiving the phenomena [of electromagnetism] was also a mathematical one, though not exhibited in the conventional form of mathematical symbols. I also found that these methods were capable of being expressed in the ordinary mathematical forms, and thus compared with those of the professed mathematicians."

Albert Einstein was a great admirer of Newton, Faraday, and Maxwell. In his office he had framed copies of portraits of these three scientists. He had this to say about Faraday and Maxwell, in "Maxwell's Influence on the Development of the Concept of Physical Reality": "The greatest change in the axiomatic basis of physics—in other words, of our conception of the structure of reality—since Newton laid the foundation of theoretical physics was brought about by Faraday's and Maxwell's work on electromagnetic phenomena...."

In 1905, just more than 100 years ago, Einstein published a series of papers that completely changed the course of physics. In these papers he made frequent and very powerful use of pictorial images (as opposed to pure mathematics), thus following in the footsteps of Michael Faraday. In later years he developed a much more formal mathematical approach to theories of gravitation and the universe. Most of the remaining years of his life were spent in searching for a formal mathematical theory that would unify all the various fields of physics in one grand theory—which would be called the *unified field theory.* He never succeeded in finding his holy grail!

In the 1950s, two physicists, Freeman Dyson and Richard Feynman, were discussing various approaches to formulating theories in physics. Feynman was seeking for unifying approaches to basic problems in physics, using visual and pictorial approaches rather than formal mathematical methods. Dyson was concentrating on a narrower range of problems because he did not believe that all-embracing theories would achieve what they were supposed to do. In his book *Disturbing The Universe,* Dyson recounts one of these discussions:

> Dick [Richard Feynman] fought back against my skepticism, arguing that Einstein had failed because he stopped thinking in concrete physical images and became a manipulator of equations. I had to admit that was true. The great discoveries of Einstein's earlier years were all based on direct physical intuition. Einstein's later unified theories failed because they were only sets of equations without physical meaning. Dick's sum-over-histories theory was in the spirit of the young Einstein, not of the old Einstein. It was solidly rooted in physical reality.

At the moment, many theoretical physicists are trying to find theories that will unify all the diverse areas of subatomic and particle physics. I believe that these theories will eventually end in failure, because they are far too complicated mathematically, and because they are not rooted in physical reality. This is why, in 1989, I started to formulate my own theory, which I described in my book *The Third Level of Reality.* In the present book I apply these ideas to some aspects of large-scale astronomy and cosmology; in particular the problems of dark matter and dark energy.

Using Faraday's Lines of Force to Explain Laboratory Physics

Here I will discuss the lines-of-force concept and use it to explain some aspects of electricity, magnetism, and light.

Magnets and Magnetic Fields

Most of us are familiar with the basic properties of magnets, either through experiments we did at school or through the use of magnets at home or work. The best-known property of magnets is their ability to attract certain metals without any physical contact, and this attractive influence can penetrate certain other materials or substances. The region of influence surrounding the magnet is called the *magnetic field.*

A good way to show the existence and extent of the field is to sprinkle iron filings on to a sheet of cardboard on top of a bar magnet. When the card is tapped gently, the iron filings will take up a certain pattern. This shows that the magnetic field acts along certain lines, and the iron filings will arrange themselves along these so-called lines of force. The lines seem to radiate from the ends of the magnet and join together halfway along the magnet. If a small compass is brought close to one end of the magnet, the compass needle will point directly away from it, whereas if brought close to the other end the needle will point directly toward it (see Figure 1.2). There appear to be two points near the ends of the magnet from which the lines radiate: These points are called the poles of the magnet. The poles can be found by placing a small compass at several points around the end of the

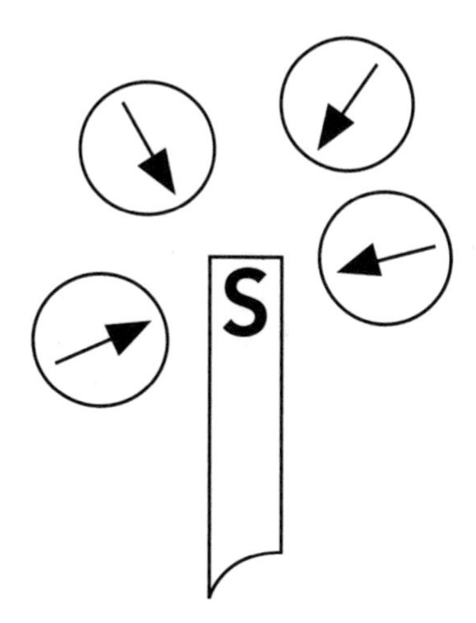

Figure 1.2
Small search compasses pointing toward the pole of a bar magnet.

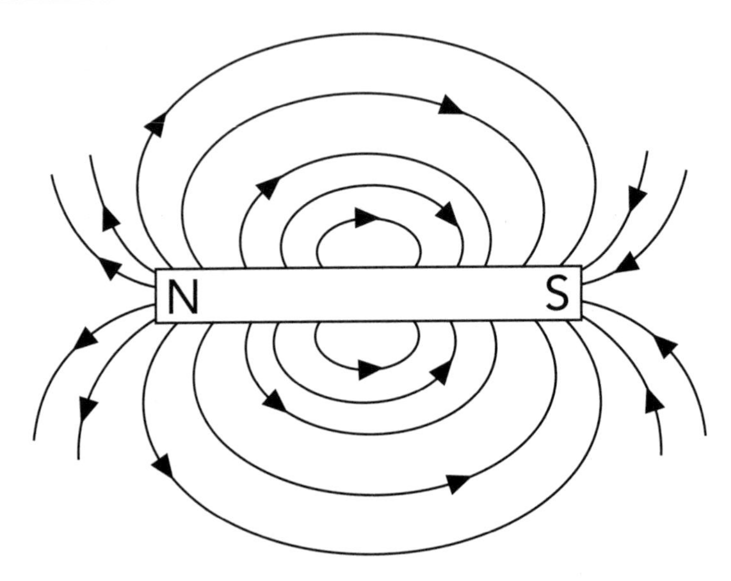

Figure 1.3
Magnetic lines of force around a bar magnet.

magnet, marking the directions in which the needle points at each position, and drawing straight lines through these directions. The point of intersection of the straight lines defines the pole at each end of the magnet (see Figure 1.3).

The magnetic lines of force are intangible lines along which a compass needle will point, or along which iron filings will align themselves if they were free to move under the influence of the magnet. Michael Faraday was able to explain a great deal by giving the lines real physical properties. He believed that these lines were like elastic bands in that they tended to contract along their lengths. However, if there was a bundle of such lines, all pointing in the same direction, then the lines would tend to repel each other at right angles to the direction in which they were pointing. In other words, there is a tension along their lengths and a pressure at right angles to their lengths. The properties of magnets can be explained in terms of the interaction between these lines of force.

Suppose that two bar magnets are marked with red paint at the ends to which a small compass needle directly points. If the two red-marked ends are brought close to each other, one magnet will repel the other. However,

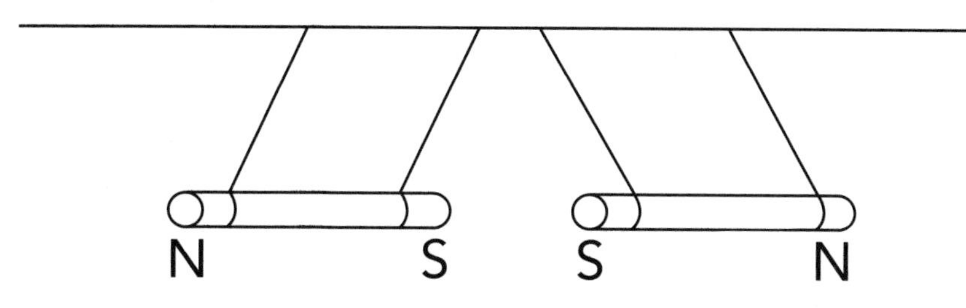

Figure 1.4(a)
Two parallel magnets, mounted end-to-end on the same rod, with like poles facing each other, will repel.

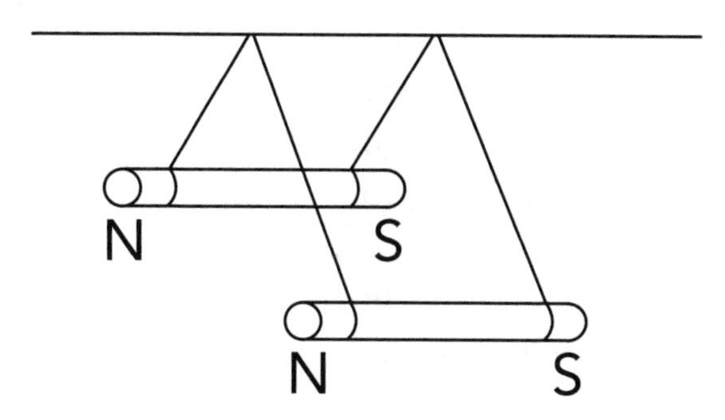

Figure 1.4(b)
Two parallel magnets, suspended on the same rod, with like poles opposite each other, will repel.

if the painted end of one is brought close to the unpainted end of the other, then they will attract each other and join together. This shows that unlike poles attract each other and like poles repel each other. It is also possible to investigate the magnetic field surrounding the magnets when they are placed in different positions with respect to each other (see Figures 1.4(a) and 1.4(b)). The magnets are stuck to a table with tape, and a sheet of card placed over them. Iron filings are then sprinkled onto the card, which is then tapped gently. The configurations taken up by the iron filings will be as indicated in Figure 1.5. In Figure 1.5(a) and 1.5(b), the lines of force between the magnets will all be in the same direction, so the pressure between the lines of force will be transferred to the magnet to which they are attached—thus the magnets will repel each other. In cases (c) and (d) the

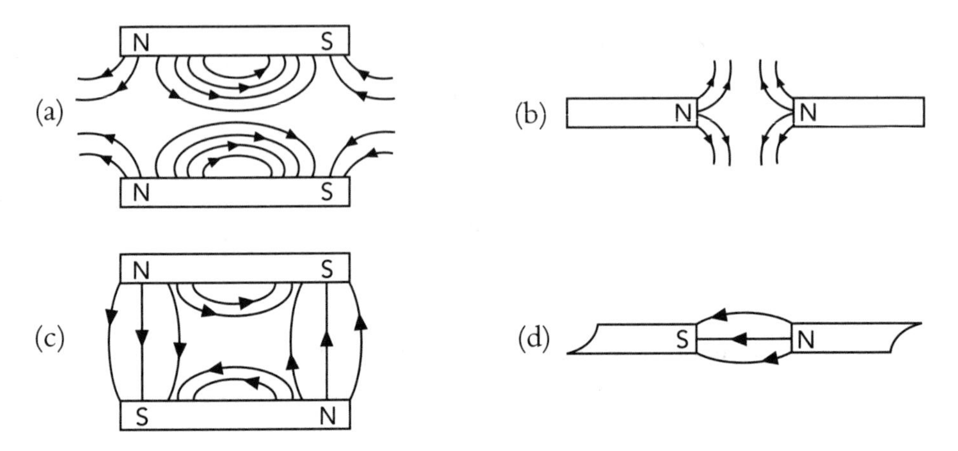

Figure 1.5
(a) Lines of force around two bar magnets with like poles opposite each other.
(b) Lines of force around the facing of like poles of two magnets.
(c) Lines of force around two bar magnets with unlike poles opposite each other.
(d) Lines of force around the unlike poles of two bar magnets.

lines will tend to contract along their lengths, so the magnets will attract each other. The explanation of the forces between magnets in terms of the lines of force surrounding them illustrates how powerful the concept of lines of force is when discussing magnetism. The same type of reasoning can be applied to the magnetic fields surrounding current-carrying conductors.

Electricity and Magnetism

Magnetic lines of force, or magnetic fields, can be generated by electric currents. If there is an electric current in a coil of wire, this will give rise to a magnetic field surrounding the coil. If the field is investigated using a small compass one will notice that it appears similar to the magnetic field of a bar magnet (see Figure 1.6). The lines of force form complete loops which thread their way through the interior of the coil. According to the modern theory of magnetic materials, it is now believed that the magnetic field lines of bar magnets also thread their way through the interiors of these magnets, and that the field itself is generated by electric currents within the atoms of the magnets.

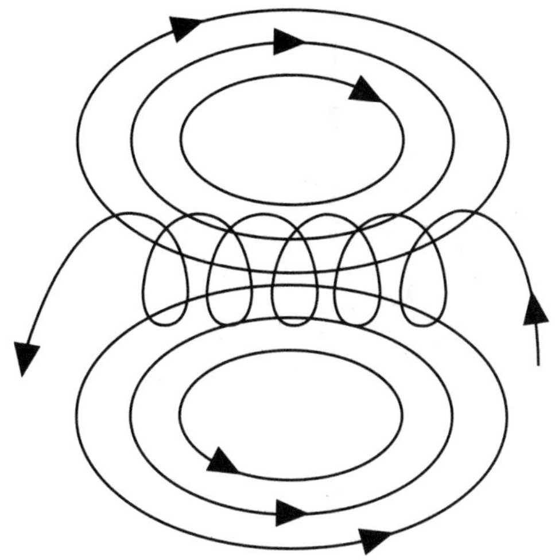

Figure 1.6
Lines of force around a coil of wire.

A straight wire carrying an electric current produces a magnetic field in which the lines of force take the form of concentric circles centred on the wire (see Figure 1.7). Two parallel wires, each carrying a current in the same direction, will be attracted to each other. This can be explained in terms of the lines of force of their combined fields. The circular lines of each individual field will cancel in the region between the wires because they are in opposite directions. There will then be a reconnection of the lines forming the rest of the circles, and the lines in the resulting loop will tend to contract along their lengths, giving rise to an attraction between the wires (see Figure 1.8).

As we have already seen, Faraday made the discovery that moving magnetic fields can produce electric currents. If a coil of wire is connected to a device for measuring electric current, called an ammeter, and a magnet is moved in and out of the coil, then the ammeter registers a fluctuating current (see Figure 1.9). In this experiment it does not really matter if the magnet is moved or the coil is moved—in either case an electric current is generated. This is the principle of the electric generator.

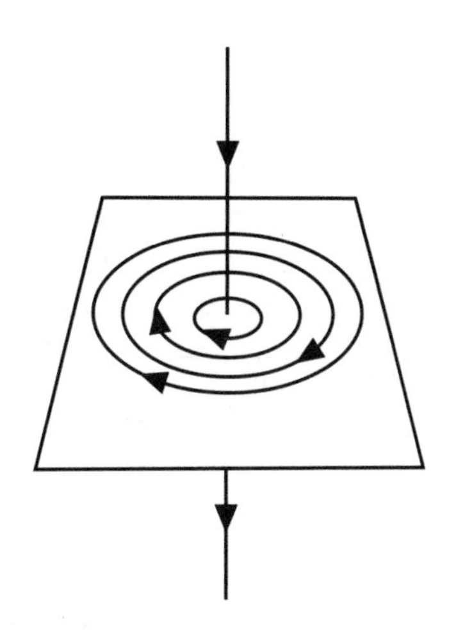

Figure 1.7
Magnetic field around a straight conductor.

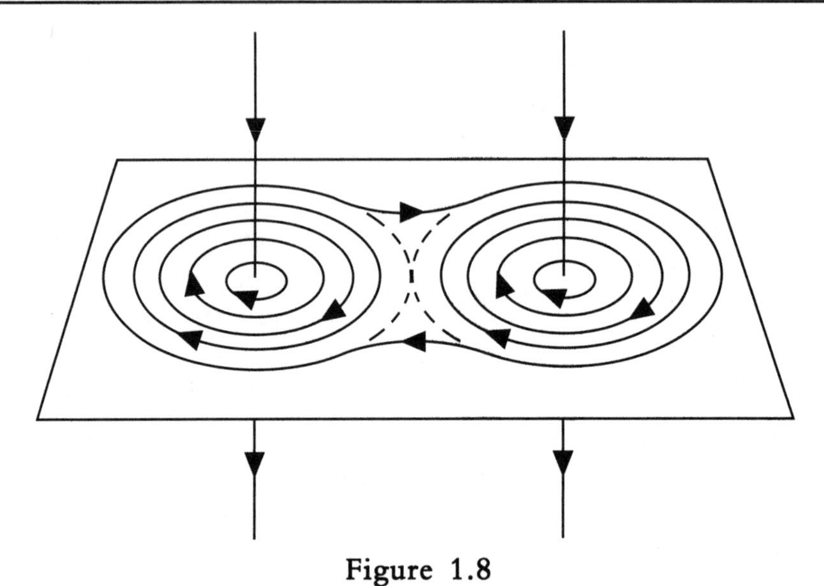

Figure 1.8
Reconnection of lines of force around two straight conductors.

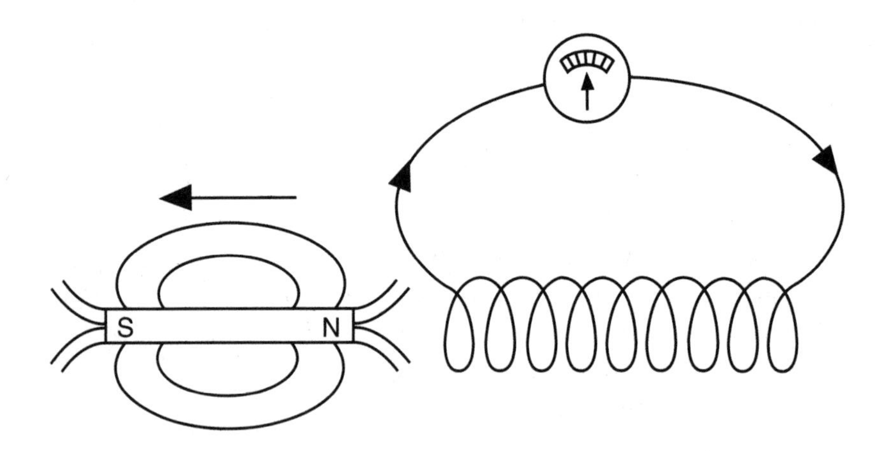

Figure 1.9
Generating electric current by moving a magnet with respect to a coil of wire.

The Motion of a Charged Particle in a Magnetic Field

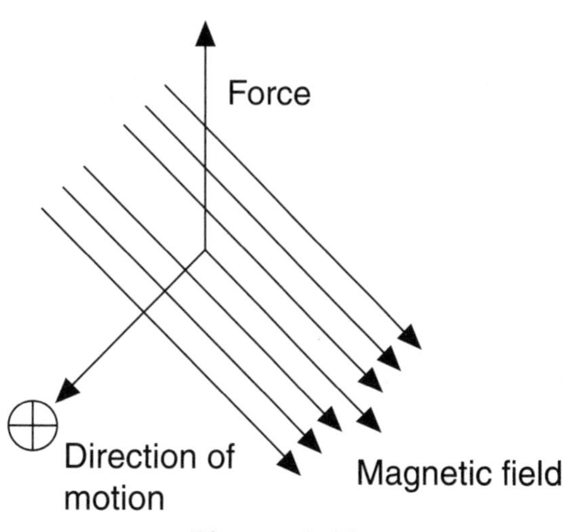

Figure 1.10

The force acting on a charged particle moving in a magnetic field.

A charged particle moving in a magnetic field experiences a force, which is at right angles to the direction of its motion, and at right angles to the magnetic field (see Figure 1.10). If the field is extensive enough and uniform in strength, and the particle's velocity is at right angles to the field lines, then its path will be a circle about the lines. If the particle's velocity makes an angle with the field lines, other than a right angle, then its path will be a helix about the lines (Figure 1.11).

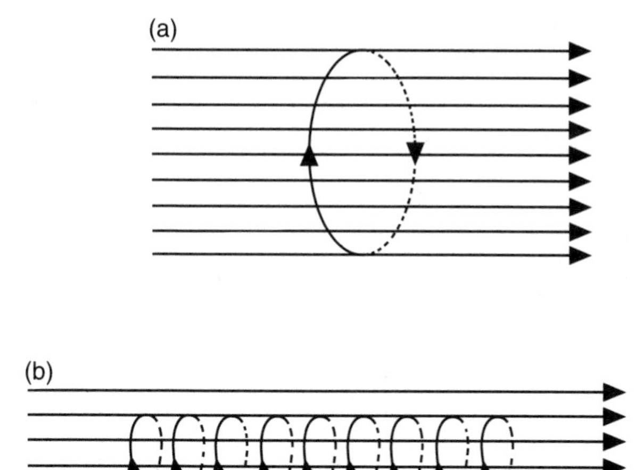

Figure 1.11

(a) Orbit of a particle with motion at right angles to lines of force.
(b) Orbit of a particle with direction of motion making an angle of less than 90 degrees with field lines.

A wire carrying an electric current consists of charges in motion within the wire, so these charges will experience a mechanical force if the wire is placed in a magnetic field. In particular, a small coil with current flowing through it, placed in a magnetic field, will have its own field, which is similar to that of a bar magnet, so it will tend to orient itself with the plane of the coil at right angles to the external field.

ELECTROMAGNETIC RADIATION

So far we have looked at the close relationship between electricity and magnetism. Now we will look at how these phenomena are linked to light and radio waves. Faraday's experiments showed that such links existed, but the mathematical genius of James Clerk Maxwell was needed to produce a theory that explained light, radio waves, X-rays, and infrared and ultraviolet radiation in terms of electrical and magnetic fluctuations. Radio waves can be generated by passing a rapidly fluctuating electric current through a straight conductor, and this is basically how radio and television transmitting aerials work. The wave coming from such a conductor would have an electrically fluctuating part parallel to the conductor and a magnetically varying part at right angles to the conductor (see Figure 1.12). In most circumstances it is the electrical part of the wave that is detected, so the wave is normally described as being polarized parallel to the conductor. The wave transmitted by such an aerial can be received by an aerial consisting of a straight conductor parallel to the length of the transmitting aerial. Some radiation will also be received if the two aerials make an angle of

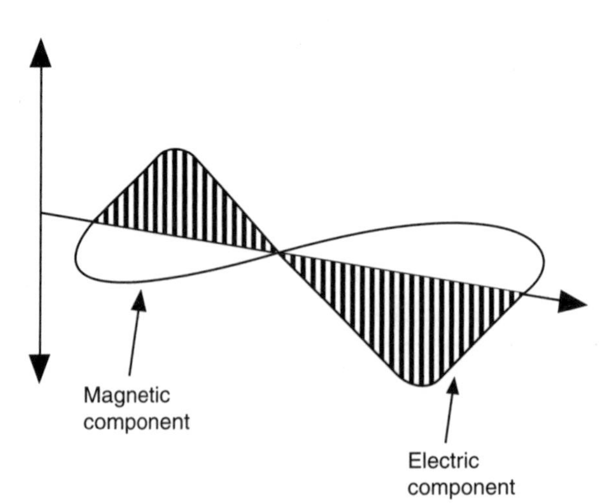

Figure 1.12
The electric and magnetic components of a plane-polarized electromagnetic wave.

less than 90 degrees with respect to each other, but hardly any radiation will be received if the angle between the aerials is 90 degrees.

LIGHT WAVES FROM ATOMS

Light waves cannot be generated in the way just described, but electrons within the atom can generate a variety of different wavelengths, from X-rays to infrared. In explaining how this is achieved, it is helpful to start with a simple picture of the hydrogen atom devised by Niels Bohr. Just as the force of gravitation of the sun on a planet is inversely proportional to the inverse square of the distance between them, so the force—due to electrostatic attraction—between a proton and an electron obeys the same law. In the Bohr model the electrons are orbiting the central nucleus, in much the same way as the planets orbit the sun, but the force of attraction is electrostatic rather than gravitational. Bohr's model of the hydrogen atom consists of one electron moving in a circular orbit about the positively charged central nucleus that has just one particle—a proton. According to the laws of classical physics, the electron would radiate electromagnetic waves as it orbits, lose energy in the process, and thus spiral into the nucleus. To avoid this, Bohr applied the ideas of quantum theory to the orbiting electron, and showed that the electron could only move in orbits at certain "allowed" distances from the nucleus (see Figure 1.13). These allowed orbits are more easily explained in terms of the full quantum theory, but such an explanation is beyond the scope of this book. However, it is possible to give a simplified explanation in terms of the ideas developed by Louis de Broglie. He assumed that, associated with any moving particle, there was a wavelength, which he called the *de Broglie wavelength*. This wavelength depends on the speed and

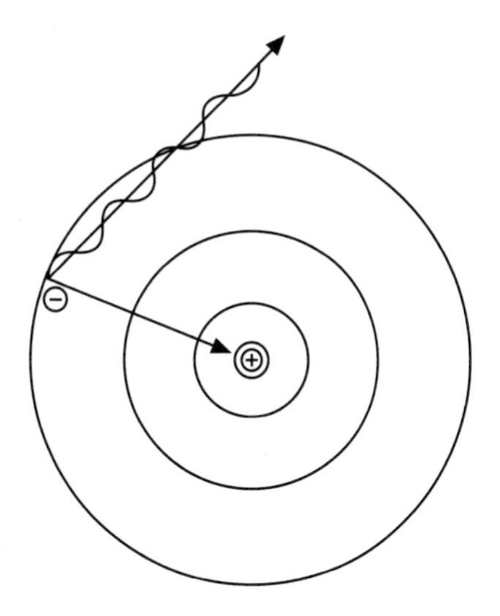

Figure 1.13
A diagram of Bohr's model of the hydrogen atom.

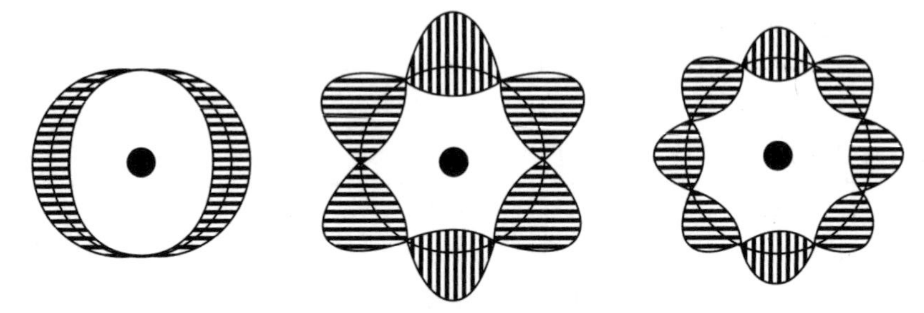

Figure 1.14
De Broglie waves and the allowed states of the Bohr atom.

mass of the particle. The allowed Bohr orbits then correspond to those in which the speed of the electron is such that the associated de Broglie wavelengths fit a whole number of times into the circumference of the orbit (see Figure 1.14).

When the atom absorbs light (or any other electromagnetic waves), the electron is shifted from an orbit close to the nucleus to one further away. Later, the electron will fall from its excited state, and when this happens, the atom will broadcast light at a precise wavelength. In other words, it will emit a narrow band of color (visible with a spectroscope), rather than the entire rainbow.

A move from each orbit far from the nucleus to one closer to the nucleus will give rise to a different wavelength, so the atom will emit a well-defined series of wavelengths characteristic of that particular type of atom. The heavier chemical elements have more complex atoms. The nuclei of these atoms consist of protons and neutrons, and a number of electrons (equal to the number of protons in the nucleus) orbiting the nucleus. Each type of atom will emit its own set of definite wavelengths. It is this set of distinctive wavelengths transmitted by the atoms of different elements that enables the chemical elements to be identified by the radiation they emit. The branch of physics dealing with this aspect of radiation is called *spectroscopy*, and it enables the chemical composition of stellar atmospheres to be deduced from an analysis of their spectra. The spectrum of a star consists of the continuous spectrum (containing all the colors of the rainbow) crossed by a number of dark absorption lines. The continuous spectrum arises from the photosphere of the star, whereas the dark lines are due to the absorption of light by the atoms of the various gases in the atmosphere of the star.

These gases can be identified by comparing the positions of the dark lines with bright emission lines generated by known elements in the laboratory.

THE DOPPLER EFFECT

A great deal of the data collected on the existence of dark matter comes from the movement of stars in galaxies, and the movements of galaxies, with respect to each other, in clusters of galaxies. Much of the information we have on dark energy comes from the speeds with which the most distant galaxies are receding from us. The most important way to measure the speeds of stars and galaxies is to use the Doppler effect, so we need to make an excursion into understanding it.

The Doppler effect was discovered in 1842 by the Austrian physicist Christian Doppler (1803–1853), in connection with his work on sound waves. One of the best known examples is the sound waves of a whistle on a fast-moving train: If you are standing on a platform as a train goes by, you first hear the train whistle as a high-pitched sound as the train approaches. At the moment the whistle passes you, the sound suddenly changes and you hear a lower-pitched sound as the train rushes by on its way. This is because sound consists of waves, which can be pictured as a series of peaks and troughs. The pitch of a given sound depends on how many peaks and troughs hit the ear in a given time—say, for example, one second. When the source of sound is traveling toward us the number of peaks striking the ear in a second is increased, and when the source is traveling away from us the number of peaks reaching the ear in a second is decreased. A diagram of this is shown in Figure 1.15.

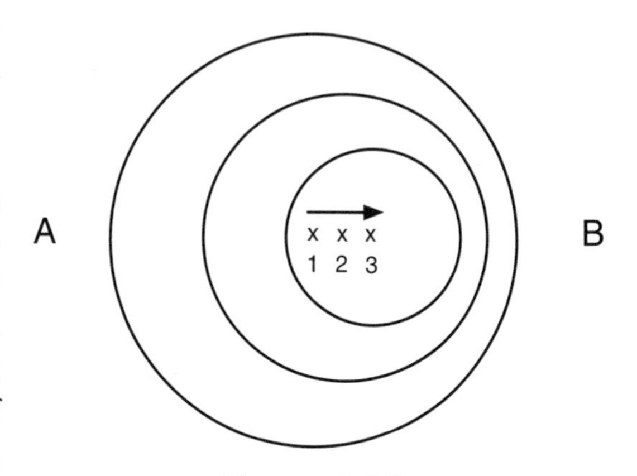

Figure 1.15
Diagram illustrating the Doppler effect. Numbers 1 to 3 are successive positions of a source of waves moving to the right.

Light waves can also be thought of as a series of peaks

and troughs, and they cause analogous effects: If a star is traveling toward us, the dark absorption lines in its spectrum will be shifted toward the blue end of the spectrum, and if the star is moving away, the lines will be shifted toward the red end of the spectrum. The amount of shift tells us the speed of approach or recession of the star.

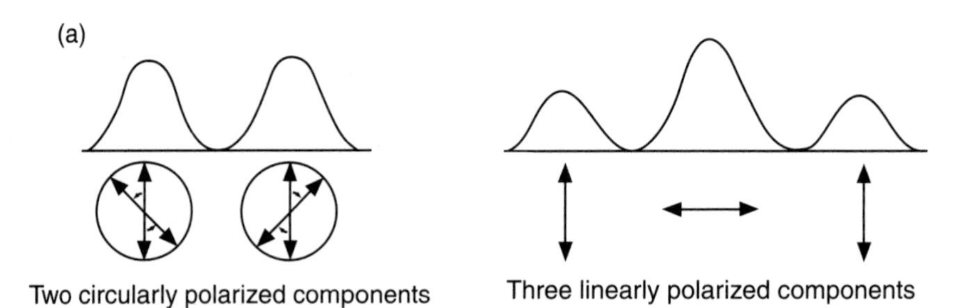

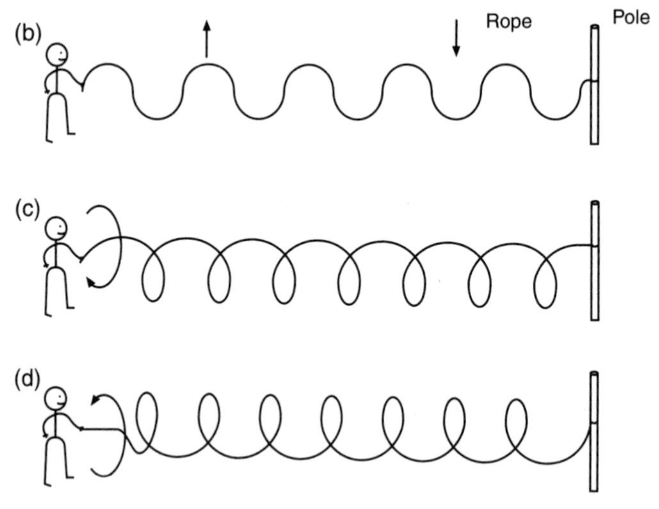

Figure 1.16
The Zeeman effect.

Measuring Celestial Magnetic Fields

A physical principle known as the Zeeman effect provides astronomers with an important tool for investigating extraterrestrial magnetic fields. In the simplest case, if the atoms of an element are placed in a magnetic field, each of the spectral lines will be split into two components when observed along the field lines, and into three components when viewed at right angles to the magnetic lines of force. The two components seen along the field lines will be right- and left-hand circularly polarized. A plane polarized wave can be seen as a rope moving up and down. Right- and left-hand circular polarization can be visualized by a rope given either a clockwise or an anti-clockwise spiralling motion (see Figure 1.16(b), (c), and (d)). The three components seen at right angles to the field lines will be linearly polarized (see Figure 1.16(a)). The wavelength differences between the components can be used to measure the strength of the field. The Zeeman effect has been used to study the magnetic activity of our sun.

The synchrotron effect is another important principle used for investigating magnetic fields in astronomy. The effect is the result of electromagnetic radiation generated by electrons, traveling at close to the speed of light, spiraling around lines of force. This effect can readily be understood in qualitative terms. According to Maxwell's electromagnetic theory, lines of force can only start and end on charged particles—they cannot originate in empty space. If a particle is at rest, an observer some distance away from the particle will see the lines of force radiating out from the point where the particle is situated. If the particle moves from point A to point B (see Figure 1.17) in a given time *t*, then the "message" that the particle has moved can only reach the observer

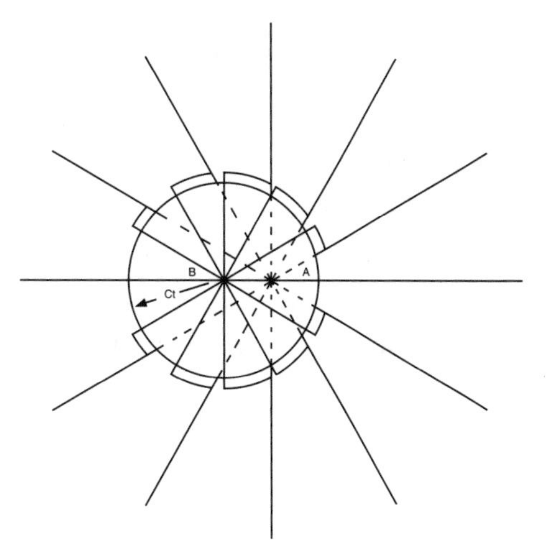

Figure 1.17
The lines of electric force radiating out from a moving point charge.

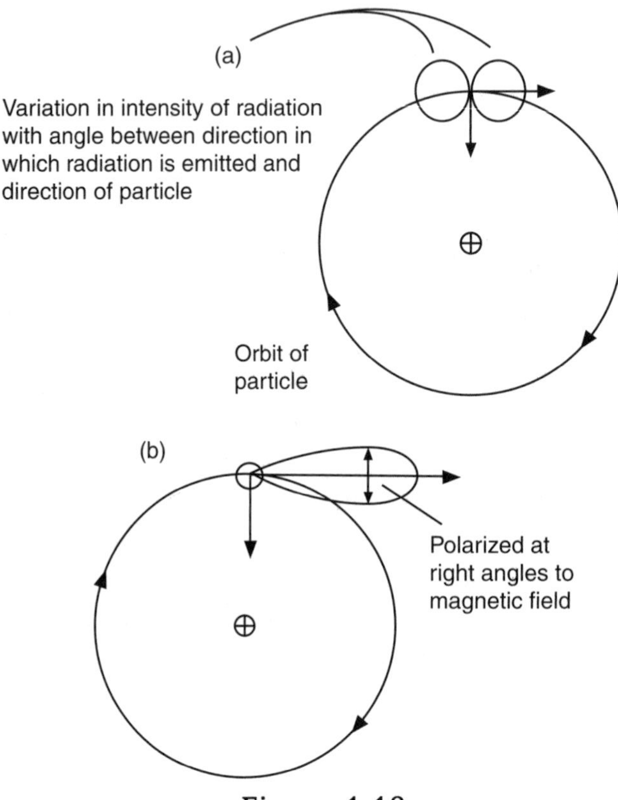

Figure 1.18

(a) Radiation from a particle moving in a magnetic field for the case when the speed of the particle is small compared with the speed of light.

(b) Radiation from a partiacle moving in a magnetic field for the case when the speed of the particle is approaching the speed of light.

with the speed of light c. So if the observer is very far from the particle, he or she still thinks the particle is at A, whereas an observer whose distance from the particle is less than ct will "see" it at the new position B. On a sphere of radius ct the lines of force seen by the two observers must join up, so there will be kinks in the lines on the surface of this sphere. These kinks are the electromagnetic waves radiating out from the particle that has moved from A to B. Any changes in the motion of the particle, whether changes in speed or in direction, can only be propagated to an observer at a speed equal to the speed of light. This means that any particle undergoing acceleration—changes in speed or direction—will have kinks in its field lines, which in turn means that it will radiate electromagnetic waves. If the speed of the particle is small compared with the speed of light, the intensity of the radiation varies with direction as compared with the direction in which the particle is moving (see Figure 1.18). If, however, the speed of the particle is very close to the speed of light, then most of the radiation will be concentrated in the region immediately in front of the direction in which the particle is moving, rather like the headlight of a motorcycle. The radiation is also polarized at right angles to the magnetic field lines if the particle is spiraling around these lines. The intensity of the radiation from a single particle is related to the strength of the field and to the energy of the particle, so measurements of the synchrotron effect can be used to investigate the structure and the strength of the field. Synchrotron radiation has been used to study the magnetic fields of Jupiter, our Milky Way galaxy, and other galaxies.

Another effect that has been used to study the magnetic field of our own galaxy is the Faraday effect. According to this principle, when a plane polarized beam of light passes through a region containing electrons and a magnetic field, the plane of polarization will be rotated about an axis that is the actual direction of the beam. The angle through which it is rotated depends on the distance over which the waves interact with the charged particles and the magnetic field, the angle between the line of sight and the magnetic

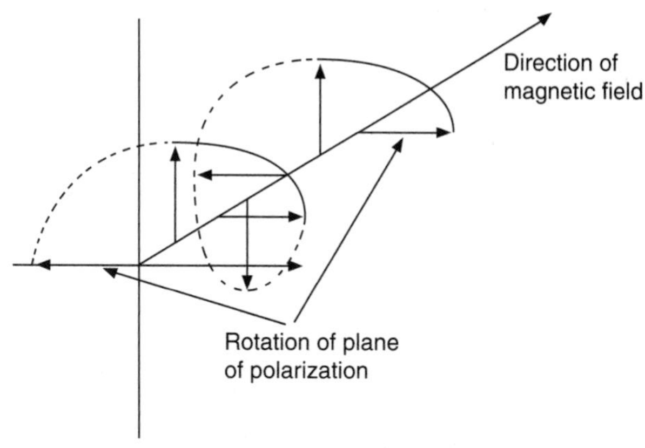

Figure 1.19

Diagram illustrating the Faraday effect for the case in which direction of propagation of the wave is along a line of force.

field, the number of electrons per unit volume of space, the strength of the magnetic field, and the wavelength being investigated (see Figure 1.19). When looking at right angles to the magnetic field there is no rotation, but when looking along the field lines the Faraday rotation is a maximum. The strength of the synchrotron radiation is a maximum when observations are made at right angles to the magnetic lines of force, and these observations are unaffected by Faraday rotation. The Faraday effect can be used to give information on the strength and direction of the field.

The magnetic fields of the Earth and planets can be measured using magnetometers onboard satellites and space probes. There are two basic types of magnetometers that have been used for this purpose. The first is called the *flux gate magnetometer*. In its basic form it consists of a core of magnetic material around which two conducting coils are wound, and in principle it is similar to an ordinary transformer. The coils are insulated from one another and from the core. A varying electric current of known frequency is passed through one coil, and the current in the other coil is analyzed by special frequency-detectors. In the presence of an external magnetic field, the core will become magnetized, and this will affect the frequencies induced in the second coil. An analysis of these frequencies will give information on the strength of the field parallel to the axis of the core. By using three separate devices such as this one, with the axes of their cores at right angles to each other, it is possible to investigate the strength and direction of the external field.

The other main type of instrument makes use of the Zeeman effect. One version of this device is known as the *alkali-vapor magnetometer*. It uses the fact that if circularly polarized light of a known wavelength is shone through the vapor of an alkali metal (such as rubidium), which is excited by a rapidly varying magnetic field, the light will be absorbed at certain frequencies related to the energy levels in the atoms, and to the strength of the external field being measured.

In the next two chapters we will extend some of the ideas from this chapter to investigate magnetic fields of our Earth, other planets, some stars, our Milky Way galaxy, and other galaxies.

Chapter Two

Magnetic Mapping of the Universe—Planets and Stars

The study of extraterrestrial magnetic fields started in the 20th century. By the end of the 19th century it was well established that the Earth had a magnetic field, and many of its properties had been investigated, including the fact that certain variations of the field were connected with the sunspot cycle: Italian astronomer Galileo Galilei (1564–1642) was among the first to use a telescope to study the sun; with it, he discovered dark patches on the sun, which we now call *sunspots*. Although they look like black dots, this is purely a contrast effect, because they are depressions in the gaseous surface of the sun that are cooler than the surrounding areas. In 1843 a German astronomer named Samuel Heinrich Schwabe (1789–1875) discovered that the number of sunspots increased to a maximum and then decreased to a minimum with a period of about 10 to 11 years. This is what is now called the *sunspot* or *solar cycle*.

Progress in the first half of the 20th century was rather slow. As we will see later in this chapter, and in more detail in Chapter 4, George Ellery

Hale, working at the Mount Wilson Observatory, had studied the magnetism of the sun in some detail in the first two decades of the 20th century. In 1909 Hale invited Harold Delos Babcock (1882–1968) to Mount Wilson to work with him, and it was here that Babcock made the first measurements of magnetic fields in stars. By 1947 the solar magnetic field had been studied in some detail, and the presence of magnetic fields in interstellar space was first inferred from the polarization of starlight, which was measured by scientists J.S. Hall and W.A. Hiltner in 1949.

In the second half of the 20th century, progress was much more rapid. This has largely been the result of three separate factors: The first is the development of the new astronomies of radio, X-ray, infrared, and ultraviolet radiation. The second factor has been the use of space probes to investigate planetary and interplanetary environments. The third factor has been vast improvements in the design of instrumentation for optical telescopes. All these factors have considerably increased our knowledge of magnetic fields in the universe. It is now clear that some planets, many stars, and several galaxies have magnetic fields that play varying roles in the structure and evolution of these objects.

In his book *Cosmical Magnetic Fields: Their Origin and Their Activity*, Professor E.N. Parker of the University of Chicago says:

> It appears that the radical element responsible for the continuing thread of cosmic unrest is the magnetic field. Magnetic fields are familiar in the laboratory, and indeed in the household, where their properties are well known; they are easily controlled, and they serve at our beck and call. In the large dimensions of the astronomical universe, however, the magnetic field takes on a role of its own, quite unlike anything in the laboratory. The magnetic field exists in the universe as an "organism" feeding on the general energy flow from stars and galaxies. The presence of a weak field causes a small amount of energy to be diverted into generating more field, and that small diversion is responsible for the restless activity in the solar system, in the galaxy, and in the universe. Over astronomical dimensions the magnetic field takes on qualitative characteristics that are unknown in the terrestrial laboratory. The cosmos becomes the laboratory, then, in which to discover

and understand the magnetic field and to apprehend its consequences.

In this chapter we will consider the magnetic fields of those planets that have such fields, and we will look at the evidence for magnetic fields in certain types of stars. We will end this chapter by looking at theories that can explain the presence of magnetic fields in these objects. However, we will start with a brief excursion into the laws of planetary motion discovered by Johannes Kepler (1571–1630); in particular, we will look at his speculations about magnetic fields in the solar system.

KEPLER'S SPECULATIONS ON MAGNETISM IN THE SOLAR SYSTEM

Central to the laws of planetary motion formulated by Kepler are the basic properties of the geometrical figure called the ellipse. The ellipse is rather like a flattened circle: A line drawn from the center of a circle to the circumference (a radius) always has the same length, no matter in which direction it is drawn. With an ellipse, this length does depend on the direction in which it is drawn. Suppose we have an ellipse with its long axis (called the major axis) in an east-west direction. A line from the center of the ellipse to the east (or to the west) will have a maximum length, whereas a line drawn from the center to the north (or south) will have a minimum length. At two points along the major axis, at equal distances from the center, are the foci of the ellipse. These foci, taken together, have an important property: If two lines are drawn from the same point on the curve of the ellipse to each of these foci, the sum of the lengths of these two lines will always be the same, no matter from which point on the curve these lines are drawn.

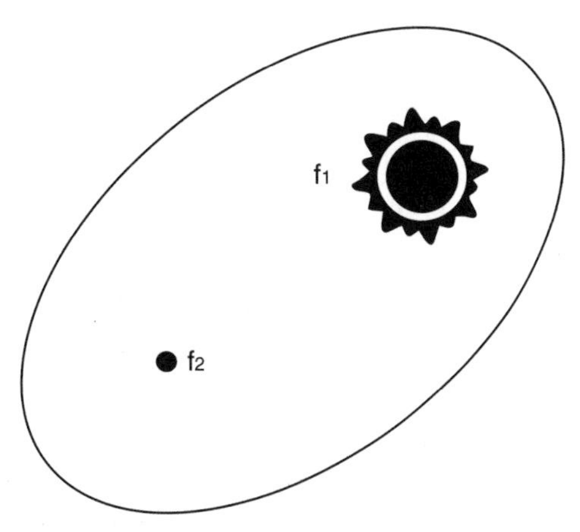

Figure 2.1
Kepler's first law.

Kepler's first law (see Figure 2.1) says that all the planets go

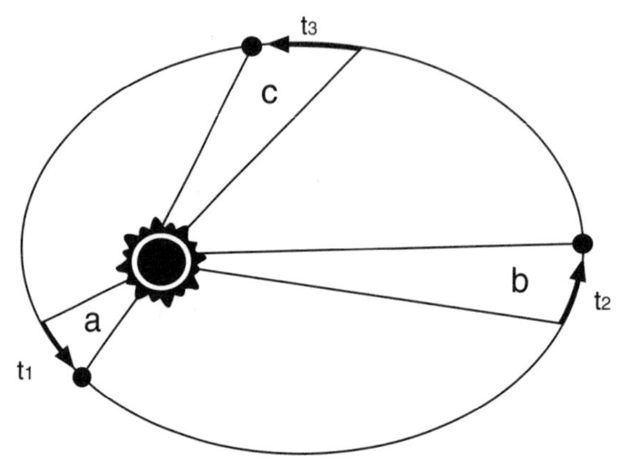

Figure 2.2
Kepler's second law.

around the sun in elliptical orbits, with the sun at one foci of each ellipse used to represent the orbit of a planet. His second law says that a line drawn from the planet to the sun sweeps out equal areas in equal times (see Figure 2.2). This means that a planet moves faster when it is closer to the sun than when it is far away from the sun. The third law states that the time it takes a planet to complete one orbit is related to its average distance from the sun. So, those planets close to the sun take a shorter time to complete one orbit than those far from the sun. For example, it takes Mercury 88 days to go once around the sun, whereas it takes Uranus 84 Earth years to orbit the sun once. The reason for this third law is not simply that the orbit is larger, but also that the planets further from the sun are moving more slowly along their orbits.

Even before Kepler formulated these laws, his physical intuition had led him to introduce forces emanating from the sun and planets to provide a causal foundation for planetary motion. The first of Kepler's forces was the *anima motrix*. According to Kepler, this was a force that pushed the planets around their orbits by a system of rays, radiating out from the sun. These rays, he believed, were restricted closely to a plane in which most of the planets moved. He reasoned that the number of rays that impinged on a planet, in a given time, as the sun rotated, would decrease as the distance between the planet and the sun increased. In his anima motrix Kepler vaguely anticipated the interplanetary magnetic field, which is carried by an energetic stream of particles coming from the sun, called the solar wind. The similarities are that they both originate on the sun, they are both nearly confined to the plane in which the planets move, and the number impinging on a planet will decrease with distance from the sun. The differences are that the lines of force of the interplanetary magnetic field are curved, it consists of four sectors, with the field pointing in the same direction in alternate sectors, and whereas the anima motrix was supposed to influence the movements of the planets, the actual interplanetary field does not have any such influence on the planets. Details of the interplanetary magnetic field will be discussed in Chapter 4.

Kepler again anticipated some modern ideas with his introduction of magnetism. In this he was influenced by the publication in 1600 of *De Magnete* by William Gilbert. Gilbert had recognized that the Earth itself was a magnet, and Kepler reasoned that not only the Earth but the sun and the other planets were also large magnets. However, the sun's magnetic field, according to Kepler, was rather unusual, in that it had it a north pole at the center and a south pole distributed over its surface. He believed that, because the magnetic fields of the planets were similar to that of Earth, with their magnetic axes nearly aligned with their rotation axes, and with a south magnetic pole near the north rotation pole, the field of the sun would tend to attract the planets along part of their orbits and repel them at other times. It was this interaction, according to Kepler, that caused the planets to move in elliptical orbits. We now know that the sun has a magnetic field, and that of the six "planets" known to Kepler, four of them have magnetic fields: Mercury, Earth, Jupiter, and Saturn.

Kepler's mathematical laws have survived for more than three centuries, but his physical ideas lasted no longer than he did; some of his ideas on dynamics were already out of date when he proposed them. For instance, the sun rotates too slowly to account for the observed periods of the planets. We also now know that the magnetic fields of those planets that have them are not strong enough, and they are not oriented in the right direction for them to do what Kepler was proposing.

The First 20th-Century Evidence for Extraterrestrial Magnetic Fields

At the end of the 19th century we had evidence that the Earth behaved like a magnet, but there was as yet no evidence for extraterrestrial magnetic fields. Some astronomers had suggested that the sun had a magnetic field, but this was just speculation based on observations that, during total eclipses of the sun, the corona (the extended atmosphere of the sun) had a vague resemblance to the patterns taken up by the iron filings around a bar magnet.

The first positive evidence that the sun had a magnetic field came from the work of American astronomer George Ellery Hale, who used the Zeeman effect to detect strong magnetic fields in sunspots in 1908. The magnetism of the sun has been studied in great detail since then, so we will leave our discussion of the magnetic activity of the sun until Chapter 4. In this section we will just mention the detection of another extraterrestrial magnetic field—that of the planet Jupiter.

In 1955, two American astronomers, Bernard Burke and Kenneth Franklin, using an array of radio aerials about 20 miles northwest of Washington, D.C., detected signals from an extraterrestrial source, which they later identified as coming from the planet Jupiter. Never before had radio signals been picked up from a planet.

More detailed radio maps made of Jupiter at 10-cm wavelengths showed that most of the radiation did not come from the planet itself, but from two lobes in either side of the planet (see Figure 2.3). Further observations showed that the intensity of the radio radiation, and its linear and circular polarization all varied with the same period. This led to the theory that attributes the radiation to the synchrotron effect operating on high-energy

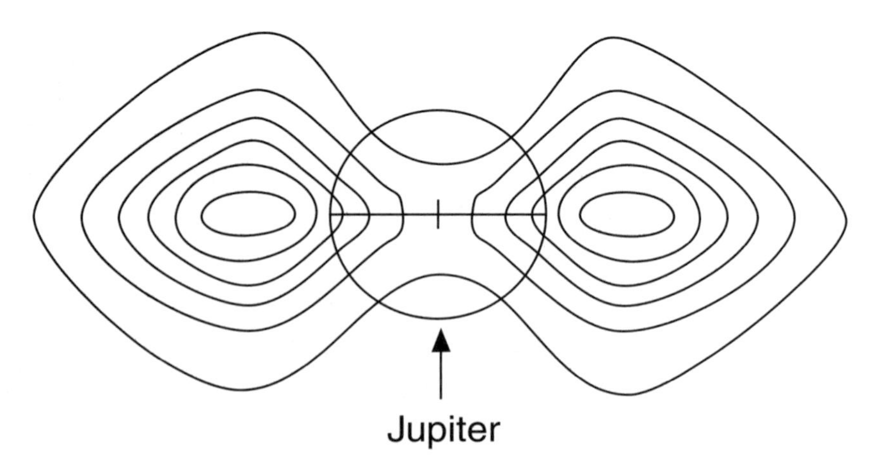

Figure 2.3
Radio map of Jupiter.

electrons spiraling in the magnetic field of Jupiter. This theory has made it possible to deduce the angle between the magnetic axis and the rotation axis. The relationships between the orientation of the magnet axis, the intensity, and polarization (both linear and circular) are shown in Figure 2.4). The angle between the magnetic axis and the rotation axis is 10.77 degrees, and the center of the bar-like "magnet" is 0.101 times the planet's radius from the center of the planet. A terrestrial compass would point to the south pole of Jupiter, and the maximum surface strength of the field is about 20 times greater than that of Earth.

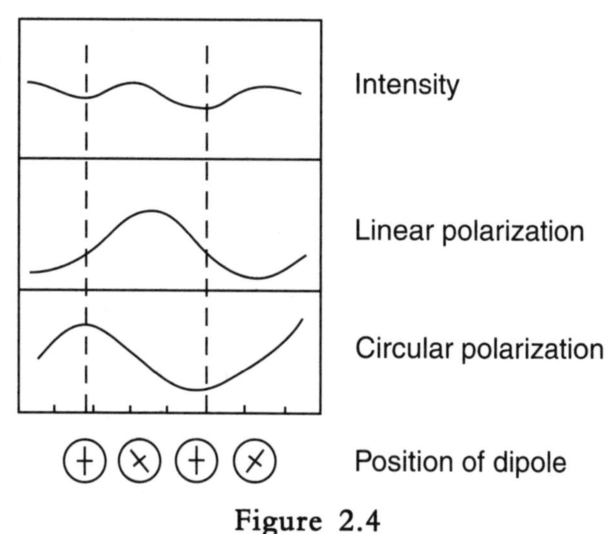

Figure 2.4
Variation of intensity, linear, and circular polarization with orientation of the Jovian dipole.

MAPPING THE MAGNETIC FIELDS OF THE PLANETS WITH SATELLITES AND SPACE PROBES

The advent of the space age opened up the possibility of investigating planetary magnetic fields, including that of Earth, in much more detail. In 1958, James Alfred van Allen (born in 1914) started using artificial satellites to explore the particle and magnetic environment high above the atmosphere of our Earth.

The Earth

The intensive research program Allen instigated led to a clarification of the structure and extent of the magnetosphere of Earth—the region surrounding the atmosphere, which was influenced by the extensive magnetic field of Earth. This magnetosphere is compressed on the side of Earth

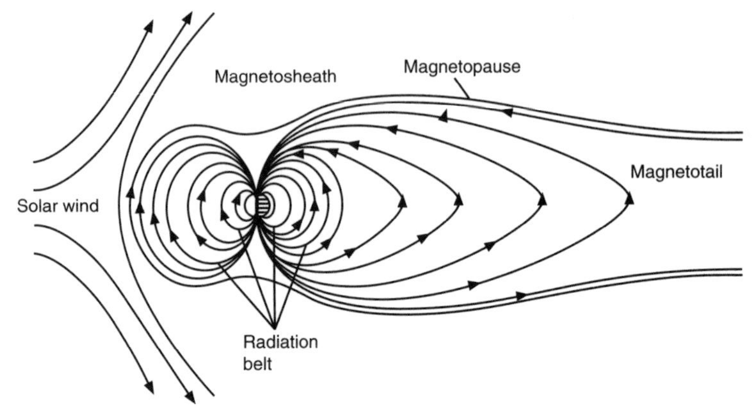

Figure 2.5
The magnetosphere of the Earth.

pointing toward the sun, and drawn out into a long tail on the side of Earth far from the sun.

The sun emits a continuous stream of energetic particles known as the solar wind. These particles consist mainly of electrons and protons, with traces of ions and the nuclei of atoms, and they are quite often referred to as *solar cosmic rays*. As they are electrically charged, they cannot readily cross the lines of force of the Earth's magnetic field, and are deflected by them. Most of the particles stream past, making a bow shock wave very similar in structure to the bow wave of a ship as it passes through the waters of the oceans. Somewhere behind the Earth the various strands of the solar wind and the interplanetary magnetic field join up again, thus enclosing the Earth's magnetic field within the pear-shaped region called the *geomagnetic tail*. Also trapped in the magnetosphere are the charged particles of the van Allen radiation belts (see Figure 2.5). Near the poles of the Earth, the magnetic field becomes much stronger, and the particles are "reflected" at these points, and tend to bounce back between these points, spiraling along the magnetic field lines. However, as a particle will follow a tighter curve in a strong magnetic field than it does in a weak one, and as the magnetic field gets weaker with distance from the Earth's surface, the path of a particle projected onto the plane of the magnetic equator will be as shown in Figure 2.6. The combination of the north-south movement and the east-west drift will cause the particles to move within the van Allen radiation belts, each belt corresponding to a different energy range for the charged particle.

The use of space probes to explore the environments of the planets have shown that other planets that have magnetic fields also have magnetospheres with similarities to Earth's.

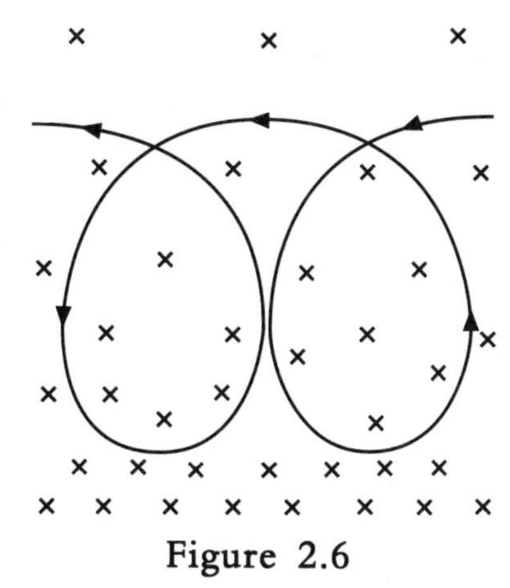

Figure 2.6
The motion of an electron in the plane of the geomagnetic equator.

Mercury

Because of its smallness compared to the other planets, and its slow rotation rate, it was predicted that Mercury would not have an intrinsic magnetic field, except, possibly, for a small field due to permanent magnetism caused by the solar wind and the interplanetary magnetic field. *Mariner* spacecraft in 1974 and 1975 measured a field with a strength of 1 percent of the Earth's field, and showed that it was dipolar in nature (it was like that of a bar magnet). Two models of the strength of the field have been proposed that lead to two different estimates of the angle between the magnetic and rotation axes: One gave a value of 2.3 degrees, and the other gave a value of 14.5 degrees. At the moment, the data we have is insufficient to give an unambiguous decision on which model is the correct one. Mercury also has a magnetosphere, although it is much smaller than that of Earth because of the weakness of the dipolar field, and, being closer to the sun, the solar wind is stronger (see Figure 2.7).

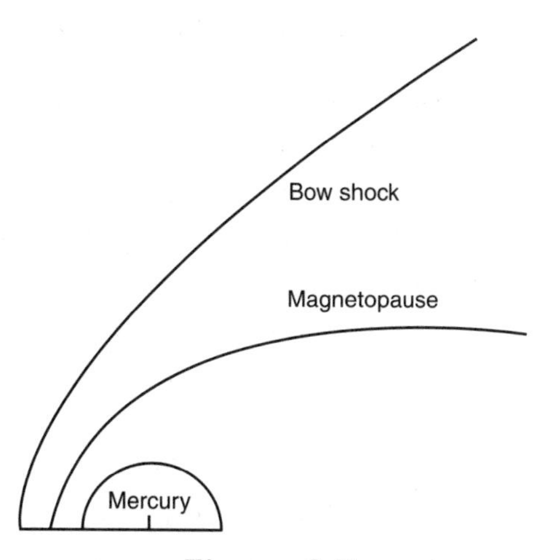

Figure 2.7
Bow shock of magnetopause of Mercury.

Venus, the Moon, and Mars

The magnetic environment of Venus was studied by a series of *Mariner* and *Veneras* space probes and also by the *Pioneer Venus Orbiter.* All these investigations showed that the strength of the dipolar field of Venus (if it indeed has one) is less than 1/10,000 of Earth's field. However, eddy currents induced in the conducting ionosphere of Venus by the solar wind prevents the wind from reaching its surface, and, as a result, the planet has a well-developed bow shock (see Figure 2.8).

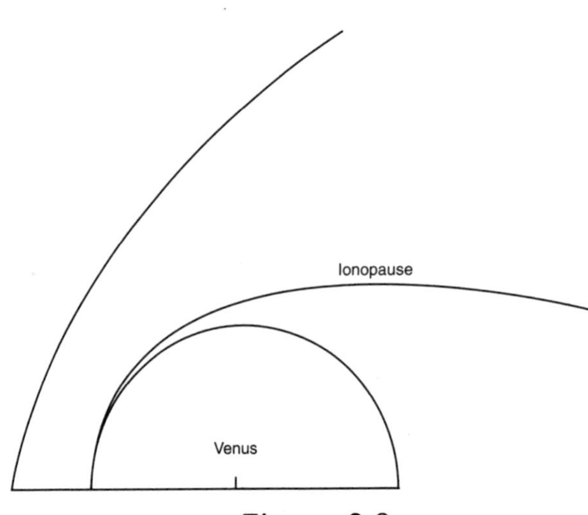

Figure 2.8
The ionosphere of Venus.

In November 2006, Lon Hood of the University of Arizona had this to say about the magnetism of our moon: "The Moon presently has no global magnetic field similar to the Earth's. The observed fields are caused by permanent magnetization of parts of the lunar crust." In my book *Cosmic Magnetism,* I said, "These studies have shown that the magnetic field of the moon is due to widespread permanent magnetism. It has also been shown that there are magnetized lava flows on the moon. However, at the moment, the moon does not appear to have an overall magnetic field of internal origin." This shows that our conclusions on the magnetism of the moon have not changed in 20 years.

As for the magnetic field of Mars, I quote from a report by Robert Sanders of the University of California at Berkeley: "Though Mars lacks a global protective magnetic shield like that of the Earth, strong localized magnetic fields embedded in the crust appear to be a significant barrier to erosion of the Mars atmosphere by the solar wind."

This conclusion was reached by a researcher at the University of California, Berkeley. It emerged from a new map of the limits of the planet's ionosphere obtained by the *Mars Global Surveyor* spacecraft, which was launched in 1996 and reached Mars 10 months later. The new data show that where the localized surface magnetic fields are strong, the ionosphere reaches to a higher altitude, indicating that the solar wind is being kept at bay.

Jupiter

The radio-astronomical observations of Jupiter's magnetosphere, which we mentioned earlier in this chapter, have been confirmed and considerably extended by data gathered by magnetometers and other equipment onboard *Pioneer 10* (1973), *Pioneer 11* (1974), *Voyager 1* (1979), and *Voyager 2* (1979). The observations have shown that the Jovian magnetosphere is the largest object in the solar system, other than the sun and its magnetic field.

The Jovian magnetosphere can be defined as the region in which the magnetic field of the planet dominates the interplanetary one. As can be seen in Figure 2.9, it consists of three distinct parts. The results obtained from the *Voyager* flybys show that the inner magnetosphere is dominated by the dipolar field out to a distance of about six Jovian radii, which corresponds with the orbit of Io (one of Jupiter's moons). In this region the magnetosphere behaves in much the same way as Earth's. Data collected by space probes has also revealed that at least one of the large satellites of Jupiter has a magnetic field, and that is Ganymede.

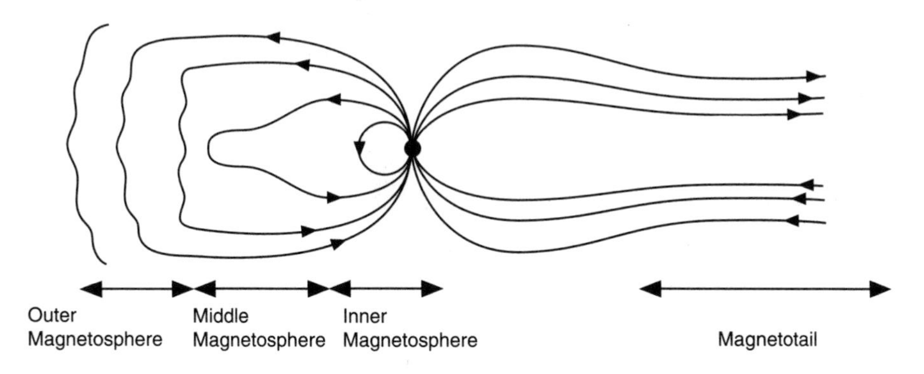

Figure 2.9
The inner, middle, and outer magnetosphere of Jupiter.

Saturn

The general structure of the magnetosphere of Saturn contains neutral gases and ionized gases (called a plasma). The two satellites, Titan and Rhea, orbit the planet within the magnetosphere. Close to the planet there is one torus (an annular distribution of material resembling the inner tire of a bicycle) consisting of molecular hydrogen gas, and farther out there is another, much larger torus, consisting of atomic hydrogen. The plasma consists of an inner sheet, between these two hydrogen toruses, and an outer sheet, which is compressed on the sunward side by the solar wind, but extends for a considerable distance on the other side.

Uranus and Neptune

Voyager 2 discovered that Uranus has a strong magnetic field. The angle between the rotation axis and the magnetic axis is 59 degrees. The dipole origin of the field is also displaced by about 30 percent of the planets radius from the center. The orientation of the planet also causes its magnetotail—the extended "tail" of the planet's magnetosphere, compressed by the solar wind on the side facing toward the sun—to be wound like a corkscrew behind the planet.

Neptune's magnetic field also does not pass through its center, and its magnetic axis is inclined by 47 degrees to its rotation axis. It is off center by about half of the radius of the planet toward the south pole. The strength of the field is about half that of Uranus, and similar to Uranus, the strength varies with latitude on the planet.

The Magnetism of the Stars

Because the sun is rather an ordinary star of average size, and possesses magnetic fields, one might expect that most similar stars should have magnetic fields. In the rest of this chapter we will look at the astronomical evidence that shows that many sun-like stars do have magnetic cycles similar to that of our own sun, and certain classes of stars have much stronger magnetic fields. The strength, detectability, and behavior of stellar magnetic fields varies from class to class, and these fields seem to be related to aspects of stellar evolution. We will begin with a brief review of stellar classification and relevant aspects of stellar evolution.

Classification of Stars

The dark absorption lines crossing the continuous spectrum of a star can be used to classify it. Originally, the varying strengths of a particular set of lines in the hydrogen spectrum, called the Balmer lines, were used to classify stars into classes labeled alphabetically from A to P; A stars had the strongest hydrogen lines. Soon after this system had been introduced, it was discovered that the spectral lines of the various elements have widely differing strengths in stars with different temperatures. As a result, a new system was introduced, which ordered the stars in a decreasing temperature sequence. This led to a rearranging of the alphabetically labeled classes and a dropping of other classes.

Most stars can be classified under one of seven spectral types, labeled by the letters O, B, A, F, G, K, and M. Three further types have been added to the scheme: W stars at the upper end of the temperature sequence, and R and N stars at the lower end. The order of this classification may be remembered using the mnemonic "Wow, Be A Fine Girl, Kiss Me Right Now!" The central mnemonic "Oh Be A Fine Girl, Kiss Me!" was first proposed by the Princeton astronomer Henry Norris Russell (1877–1957), and still remains popular, but the rise of the feminist movement has led to alternatives, one of which is: "Only Boys Accepting Feminism Get Kissed Meaningfully."

These spectral types can be further subdivided into 10 subclasses, denoted by the numbers 0 through 9, placed after the spectral type letter. Special spectral characteristics within a subclass are indicated by an additional small letter, placed after the spectral type.

Stars also vary a great deal in absolute brightness, or, to use a more objectively measurable quantity, luminosity. The luminosity is defined as the total energy radiated per second from the star. The Hertzsprung-Russell diagram (see Figure 2.10) shows the relationship between the spectral type and luminosity of stars. The brightest stars are also the hottest stars. The luminosity of a star is also related to its mass, more massive stars being more luminous than less massive stars.

The internal structure of a star is related to its total mass. For our sun, and stars of similar or lower masses, about 1 percent or more of the mass is in the form of convective envelopes (which occur when convection occurs in the outermost shells of a star). For stars between 15 and 1.5 solar masses, between

38 percent and 6 percent of the mass is in the form of convective cores (when convection occurs in shells closer to the center of a star). As we will see at the end of this chapter, convection very likely plays an important part in one of the theories as to how stars generate their magnetic fields. The fact that all main sequence stars (those lying along the curve shown in Figure 2.10) have convective regions, provides an important reason for believing that many stars should have magnetic fields at least as strong as that of our sun.

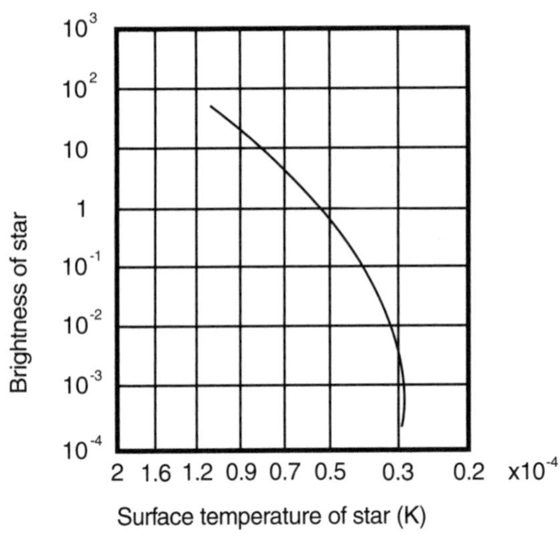

Figure 2.10
The Hertzsprung-Russell diagram.

ASPECTS OF STELLAR EVOLUTION

Stars are formed out of the collapse, under the force of gravitation, of vast gas and dust clouds that exist between the stars. As such a cloud (many hundreds or thousands of times the mass of our sun) contracts, it will fragment into smaller clouds, with masses that can range from fractions of the sun's mass to masses of a few times the mass of the sun. As these smaller clouds collapse even further, the temperature, pressure, and density of the center of the cloud will increase considerably, until the temperature is high enough for nuclear reactions to occur. At this stage the proto-star becomes a main sequence star, and most stars will spend the major part of their evolution in this phase. During this stage the star is mainly fueled by converting hydrogen into helium in the core, although other nuclear reactions occur in some of the heavier stars.

Once all the hydrogen in the core has been converted into helium, the energy generation required to prevent the star from collapsing under the force of gravitation ceases. At this stage, the core collapses until sufficiently high temperatures are reached to burn hydrogen in a shell around the now-inactive helium core. This is the shell-burning phase. The outer envelope of the star expands considerably, and cools as it does so. As a result, the star will move off the main sequence to become a red giant. Although the outer envelope of the red giant expands considerably and becomes less dense, the core itself contracts a little and becomes more dense. As the density increases, the matter in the core assumes a new state, called *degenerate matter*.

Degeneracy is a quantum mechanical effect, and in order to understand the concept it is necessary to explain two fundamental principles of quantum mechanics. The first is called the *Pauli exclusion principle*. According to this principle, only two electrons, spinning about their own axes in opposite directions, are allowed to exist in a very small volume of space, called *phase space*. The size of the volume of phase space is determined by Heisenberg's uncertainty principle. According to this principle it is impossible to know the position and the speed (or, more precisely, the momentum) of a particle with the same accuracy at a given time. In other words, if we know the position of a particle with great accuracy, then we cannot know much about its momentum. As a consequence of this law, if an electron is confined to a small region of ordinary space, its position will be known with great accuracy, but little will be known about its momentum—it will be moving about in all directions, at high speed, rather like a caged animal. As a result of this random motion, the electron exerts a pressure on the walls of its container. This pressure is called *degenerate pressure*. It is this pressure that prevents the core of a red giant from collapsing.

In the red giant phase, hydrogen burning (the conversion of hydrogen into helium) continues in spherical shells further and further from the center, and the resultant helium is added to the core. The degenerate matter in the core contracts still further, and heats up until a temperature of 100 million Kelvin is reached. At this point, helium burning begins, in which helium is converted to heavier chemical elements, such as carbon. Successive nuclear fuels are used up until iron is created. Making elements heavier than iron from those lighter than iron does not yield energy, but uses energy! When all the available nuclear fuels have been used up, there is no energy to support the star against the force of gravitation. At this stage the star will contract considerably to become a white dwarf.

Most of the matter in the core of a star that collapses to become a white dwarf is degenerate, with the electrons stripped from their nuclei and all the particles packed very tightly together, so the internal pressure comes from the degeneracy of the electrons. The star will continue to contract until the attractive forces due to gravity are balanced by the degeneracy pressure. Theoretical calculations on the structure of white dwarfs show that such stars cannot have a mass exceeding 1.44 solar masses. This does not mean that all stars above this limit cannot form a white dwarf. It is possible for a collapsing star to shed some of its outer shell, and so, provided that the matter in the core does not exceed 1.44 solar masses, the star may still end up as a white dwarf. If the amount of material in the core does exceed this value, then it is quite likely that still further collapse will occur and the star will end up as a neutron star or as a black hole.

Neutron stars are believed to be formed when the matter in the core of the collapsing star exceeds 1.44 solar masses (this limit is called the *Chandrasekhar limit*, after the astronomer who first worked out the theory of white dwarf stars). If the core mass exceeds this value, then the force of gravitation is greater than the degenerate pressure, and the resulting collapse will further crush the electrons into the nuclei to form neutrons. Neutron stars are prevented from further collapse by degenerate neutron pressure. However, if the mass of the collapsing core exceeds about 3 solar masses, then the force of gravitation is even stronger, and the degenerate neutron pressure is not sufficient to support yet further collapse to form a black hole.

A collapsing star becomes a black hole when its radius has shrunk below a certain limit known as the *Schwarzschild radius*. Below this limit the surface gravity is so high that no radiation originating in the sphere with this radius can escape from this region. This spherical surface is called the *event horizon*, and it defines the boundary inside which all information is trapped. The spherical mass will continue to contract, because a black body does not really have a stable state. However, the event horizon has a fixed radius with respect to the center of the mass, so the surface of the black hole continues to recede from the event horizon, as it contracts.

Although the total mass of a star is an important contributory factor in determining the final evolution of a star, the details are by no means clear. Theories that attempt to deal with steady mass loss from stars, catastrophic mass ejection, and supernova explosions, are still at very early stages.

Stars with less than 1 solar mass have lifetimes that are longer than the present age of the universe. If the mass of a star is greater than or equal to

1 solar mass, but less than 1.44 solar masses, the star is likely to end up as a white dwarf. If the mass of a star is greater than 1.44 solar masses, but less than 3 solar masses, it may end up as a neutron star, or, by ejecting mass so that the remaining core is less than 1.44 solar masses, it may end up as a white dwarf. For stars greater than 3 solar masses the situation is much more unclear. They could either end up as black holes, or by ejecting mass, as neutron stars. More theoretical research is needed before these questions can be resolved.

Detecting Magnetic Fields of Stars

The great distances to other stars, compared with the distance to our sun, presents a major difficulty in the investigation of stellar magnetic fields. As we will see in Chapter 4, the magnetic polarity fluctuates over the visible disc of the sun. Sunspots tend to come in pairs, and each such pair is often linked by an arch of hot gas, called a *loop prominence*. The magnetic field in sunspot pairs has different polarities in each half of the pair: The dipole field, which is most obvious near the poles, is naturally of opposite polarities; the general field varies in polarity from one region to the next. However, because of the closeness of the sun, it is possible, with even moderately sized telescopes, to study the different parts of the sun separately. This cannot be done with most other stars, for which it is only possible to study the combined light of each star. The effect of different polarities will tend to cancel each other out, and so no net effect will be measurable by the Zeeman splitting of spectral lines in sun-like stars. The application of the Zeeman effect to the spectra of certain classes of stars has shown that some of them do possess magnetic fields much stronger and more ordered than that of the sun.

Many of the strongly magnetic stars are a special subclass of A-type stars, called *A peculiar* or *Ap stars*. The proportions of the different chemical elements that exist in a given class are more or less the same. However, in the Ap stars, certain elements, such as manganese, silicon, strontium, and the rare earths, are more abundant than in normal A-type stars. Other peculiarities of these stars are variations in their brightnesses and changes in their spectra.

Some Ap stars are rotating rapidly (similar to many other kinds of stars). Because of the Doppler effect, it is difficult to detect the Zeeman splitting

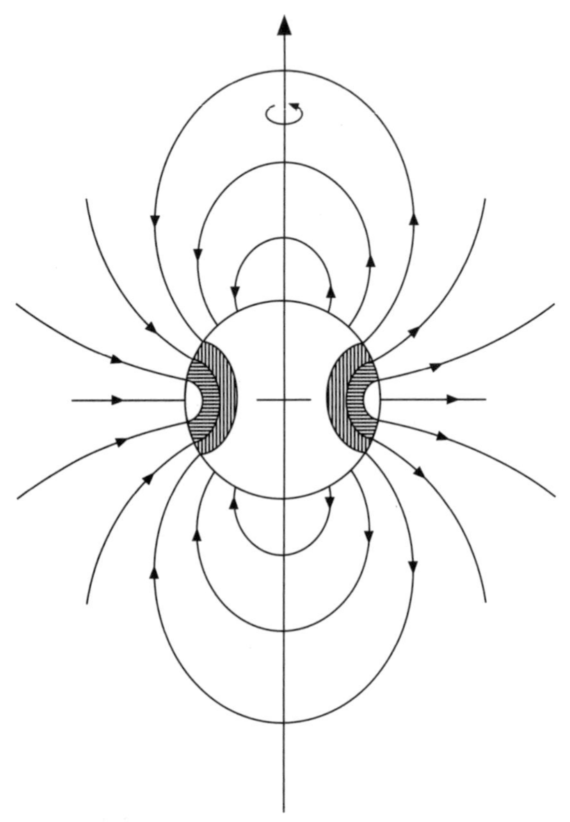

Figure 2.11
Model of an Ap magnetic star with dipolar axis at right angles to the rotation axis.

of the lines in the spectra of rapidly rotating stars: As a star rotates, part of it will be traveling toward us, and part of it away from us; as we are studying the combined light from the whole star, the spectral lines will be broadened by this rotation. It is this broadening that makes it difficult to detect the slight splitting that may be present due to the magnetic field of a star. Astronomers have been able to measure the magnetic fields in most Ap stars that are not rotating fast enough for the Doppler broadening to mask the Zeeman splitting of the spectral lines. These measurements show that the magnetic field being measured changes periodically, and the period of change, in most cases, is also reflected in other periodic changes in the spectra and light output of these stars. The observations can be understood in terms of a rather simple model, in which we have a dipole field with an axis at right angles to the axis of rotation (see Figure 2.11). The magnetic field near the poles of the dipole will affect the rate at which certain chemical elements diffuse upward to the surface of the star, so the concentration of these elements will be different at the poles than elsewhere on the surface of the star. This concentration will not only affect the observed spectrum of the star, but also its light output. Thus as the star rotates, different parts of the star will point toward the observer, and the measured magnetic field, light output, and spectrum will vary with the same period.

This model is not the only one that has been proposed. Another possibility has been suggested, called the *oblique rotator model*, in which the magnetic dipole is not aligned along the rotation axis, but is inclined to it at an angle of less than 90 degrees. Both models are able to explain the observations, and the question of which is the correct one depends on which theory for the origin of the magnetic field turns out to be the more acceptable. The two main theories will be discussed at the end of this chapter.

Magnetic White Dwarf Stars

In the last few decades magnetic fields have also been detected in white dwarfs, and the number of these having such fields is growing. The main method of detection has been the Zeeman effect. However, the presence of magnetic fields in other white dwarfs has been deduced because of the polarization of their radiation. The stars are believed to have dipolar magnetic fields, with strengths between 100 and 10,000 times more than that of magnetic Ap stars.

Magnetic Fields in Sun-Like Stars

Scientists have known for some time that emission lines due to calcium are enhanced in the spectrum of those regions of the sun, such as sunspots, where the magnetic field is strong. On the basis of these observations, astronomers at Mount Wilson Observatory in Southern California began a search for fluctuations in the calcium emission lines of nearby sun-like stars. Throughout a period of 14 years they detected such variations in a number of stars. From this they concluded that some 91 stars similar to our sun undergo cycles similar to the 22-year sunspot cycle of the sun. The proportion of the stars they studied that turned out to be magnetic led these astronomers to argue that about half of the stars similar to the sun have magnetic fields.

Since this work at Mount Wilson there have been several other investigations into stellar magnetic activity using different "activity indicators." From studies of our sun it has long been obvious that solar magnetic activity is associated with X-ray and ultraviolet emission from the corona. By analogy, it would be expected that other sun-like stars, for which it is difficult to measure magnetic fields directly, would nevertheless exhibit the associated X-ray and ultraviolet output. The *International Ultraviolet Explorer* and the *Einstein Observatory* satellites have been used to study different types of stars to search for these effects, and the results have been spectacular, leading astronomers to believe that magnetic activity in certain stellar types is fairly widespread.

The total light outputs from a certain type of star called *flare stars* show variations that can only be explained in terms of a large, cool "sunspot" rotating with the star, so the star is appreciably fainter when the spot is facing toward the telescope. Some of these stars are less bright than the sun, but they frequently exhibit flares a thousand times larger than the largest observed on the sun. In these cases the whole star will be brightened by the flare, and this is why it is possible to observe these flares from very great distances. The "starspots" can be used to study stellar activity in these types of stars.

PULSARS

In 1968, Antony Hewish, together with some radio-astronomers at Cambridge, announced the discovery of rapidly pulsating radio sources (pulsars) with incredibly stable periods. The actual discovery was really made by Jocelyn Bell (now Bell-Burnell), a postgraduate student working with Hewish, but her initial discovery was followed up by the whole team. This rapid pulsation presented a severe problem for theorists, because it was difficult to explain the pulses in terms of the vibrations of a white dwarf star. Although the existence of neutron stars had been proposed by two astronomers, Walter Baade and Fritz Zwicky, in 1934, it was also difficult to see how neutron stars could pulsate with these periods. The late Thomas Gold (proposer, with Herman Bondi and Fred Hoyle, of the steady state theory of cosmology) suggested that pulsars were rapidly rotating neutron stars, but it was some time before this suggestion was developed into a theory able to explain the growing number of observations.

Observations of the first pulsar were made at radio wavelengths of a few meters, and these revealed a pulse period of about 1.3 seconds. Although the pulse rate was steady, the amplitude varied from one pulse to the next. The arrival time for a given pulse varied with frequency, lower-frequency pulses arriving at a later time than their higher-frequency equivalents. This delay is due to the fact that the interstellar medium has a refractive index that is related to frequency—waves of different frequencies travel at different speeds. This delay can be used to estimate the distances to pulsars. (This subject will be discussed more fully in the next chapter.)

The pulses are also all polarized, although the plane of polarization is different at different times, even within a given pulse. The angle of polarization is frequency-dependent, showing that Faraday rotation is occurring in the interstellar medium. Also in the next chapter we will see how the Faraday rotation measures of pulsars can be used to probe the magnetic field of our Milky Way galaxy.

Although at first the lengths of pulsar periods seemed to be constant, further investigations showed that they increased with time. The rate of increase of period with time has been measured for several pulsars. Another property of the period of some pulsars is that they occasionally take a sudden increase, followed by the normal pattern of steady increase with time. These sudden jumps are called *glitches*, and they are believed to be starquakes due to sudden releases of strain in the crust of the neutron star.

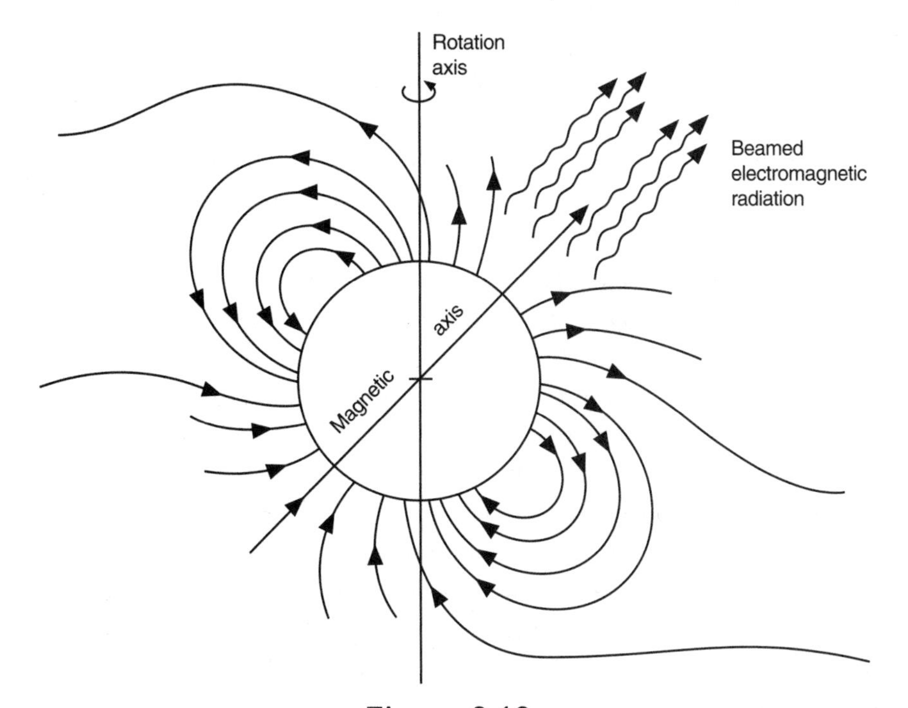

Figure 2.12
Magnetic field of a neutron star.

Most of the pulsar's emitted radiation comes from its magnetosphere. It is now generally accepted that the neutron star at the center of the pulsar has a dipole magnetic field that is not aligned with the rotation axis (see Figure 2.12). Electrons spiraling in the magnetic field produce some of the radiation by the synchrotron effect. However, near the poles, where the field is particularly strong, some of the electromagnetic waves arise from curvature radiation. In this process the helices of the electron orbits are so tight (that is, the radii of curvature so small) that the electrons virtually follow the field lines. Because the field lines are generally curved, any electron will be undergoing acceleration toward the center of curvature, and will radiate in the direction in which it is moving (just as for synchrotron radiation) because its speed is close to the speed of light. The radiation from either process will be particularly strong at the poles, and this radiation will be beamed outward. As the star rotates, the beam will rapidly sweep across the radio telescope, rather like the beam from a lighthouse, and this will be interpreted as a pulse of radio waves. Because of the intense magnetic field and the high energy of the electrons, pulsars would be expected to send off pulses of light waves, in addition to the radio waves they are known to emit, and such pulses have been observed from the pulsars associated with the Crab and Vela supernova remnants. Some pulsars also send off X-rays, but this radiation is believed to be due to the accretion of matter onto the magnetic poles where the matter has been channelled by the magnetic field.

The pulsar in the Crab Nebula has received special attention. The nebula itself has been the subject of intense astronomical research for a number of years, and it is a particularly interesting example of a supernova remnant. The supernova explosion that gave rise to the Crab Nebula was first seen by Chinese astronomers in AD 1054. Throughout the last 50 to 60 years it has been studied by optical, radio, X-ray, and theoretical astronomers, and at least some of the problems associated with it have been solved.

Radiation from the Crab Nebula is highly polarized at both optical and radio wavelengths. Astronomers now believe that almost all of its radiation is generated by the synchrotron process. The nebula is able to radiate by this process in such a wide range of wavelengths because the field is very strong, and the charged particles are very energetic. We have already seen that this type of radiation is plane polarized at right angles to the magnetic lines of force, and so it can be used to give us information on the direction of the magnetic field in different parts of the nebula.

Magnetic Fields in Double-Star Systems

Ever since their discovery, double-star (or binary) systems have played an important part in astronomy. They are the only stars for which it is possible to determine the mass with some precision. It now seems likely that they hold important clues to stellar magnetism. In some such systems there are observed light variations different from the orbital periods. These can be explained in terms of the presence, on one of the stars, of cool, large regions (starspots) rotating with one of the stars. This would seem to indicate the presence of magnetic fields.

For one such type of star (called an *RS Canum Venaticorum* type), the presence of magnetic fields is further supported by flaring, polarized radio emissions, and strong X-ray and ultraviolet emissions. More direct evidence has recently come from spectroscopic studies of these systems. For example, certain spectral lines are more sensitive to Zeeman splitting than others. By comparing those sensitive to Zeeman splitting with those that are less sensitive, it has been possible to infer fairly strong magnetic fields in at least one such system. Studies of these stars have thrown light on the possible relationship between stellar age and activity. In recent years it has become clear that age seems to affect stellar activity, mainly in that stars lose their angular momentum (their spin rate slows down) as they age. The RS Canum Venaticorum stars show high activity in spite of their age. This can be explained by attributing the maintenance of the rotation to the tidal effects between the two components of the system.

Magnetic Fields in X-Ray Sources

X-ray astronomers have discovered that, generally, the strongest point-like sources of X-ray radiation in our galaxy are binary systems in which mass is being transferred from one component to the other. Magnetic fields also play a part in such systems. The X-ray source Hercules X-1 is fairly typical of such systems; it consists of a neutron star with a spin period of 1.24 seconds, which orbits a companion with a period of 1.7 days. This orbital motion leads to the neutron star being eclipsed by its companion, but, because of the Doppler effect, it also leads to variations in the observed period of the neutron star. A diagram of the model proposed to explain this source is shown in Figure 2.13. The neutron star is accreting matter from the normal star H2 Herculis. The incoming gas will be funnelled to the polar regions of the star's strong magnetic field (see Figure 2.14). The energy gained by the gas as it falls toward the compact neutron star will cause it to heat up to a very high temperature; consequently, the gas will emit X-rays. Some of the X-rays from the neutron star will be intercepted by the normal star, thereby causing an increase in the temperature of that side of H2, which points toward Hercules X-1. This increase in temperature is observable in the light received from the system.

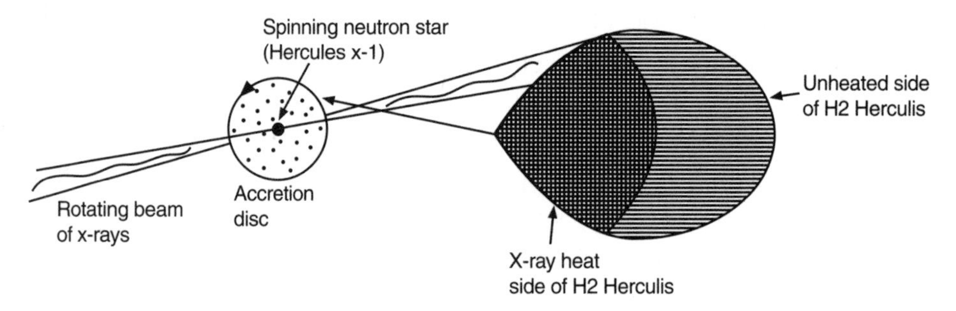

Figure 2.13
Model for the Hercules X-1/H2 Herculis system.

THE ORIGINS OF PLANETARY AND STELLAR MAGNETIC FIELDS

Magnetic fields are not an intrinsic property of space or of matter. This means that whenever we encounter magnetic fields, whether it is in the laboratory, in rocks, in planets, or in stars, we look for some mechanism that could have given rise to them.

The generally accepted theory for the generation of the magnetic fields of the planets is called *dynamo theory*. Essential to an understanding of dynamo theory is the concept of the freezing of magnetic lines of force into electrically conducting materials. The conducting medium can be either molten metal, or plasma, which is a highly ionized gas. "Freezing in" means that if one moves the conductor, then the magnetic lines of force will follow

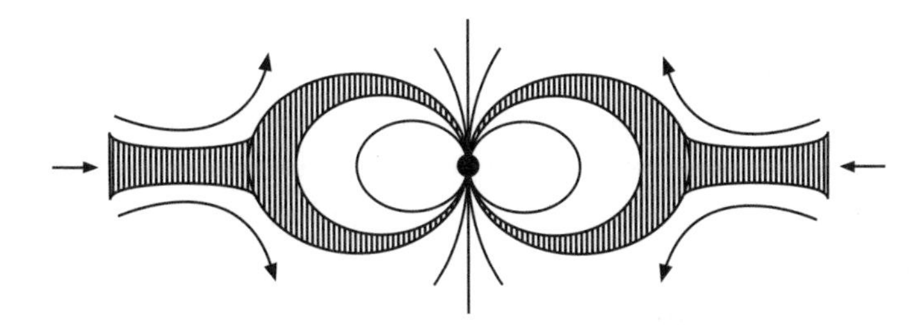

Figure 2.14
Accretion of matter by a neutron star with a strong magnetic field.

this movement, and if the magnetic field is moved, then the conductor will also follow this movement.

The Earth's magnetic field is generated in the conducting material of its fluid outer core. Initially the dipole field will pass through this core, going from the north magnetic pole to the south magnetic pole. The outer core, which lies beneath the crust and the mantle, is rotating. However, it does not rotate as a solid body. The rotation rate is latitude dependant—the material closer to the equatorial plane of Earth rotates faster than those regions closer to the poles. Thus, a single line of magnetic field, originally going from pole to pole, will be gradually distorted. Convection currents (similar to those observed in soup when it is heated on a domestic cooker) in the outer regions of the outer core will then twist loops into the magnetic field, and these loops will then combine to give rise to a new field with the same polarity as the original field. If such a mechanism did not exist, the Earth's magnetic field would have decayed a long time ago. This is because of Faraday's assertion that there is a magnetic pressure at right angles to a bundle of lines of force all pointing in the same direction. This pressure would have expelled the old field throughout a short period of time, compared with the age of the Earth. The same will happen to a single bar magnet in the laboratory, which is why bar magnets are stored in pairs with a strip of iron over the poles. The field lines will then be in the form of complete loops, and the magnetic pressure will be balanced by the magnetic tension in the loops. Similar methods are most probably at work in the interiors of those planets that have magnetic fields.

The next question that has to be answered, with respect to the Earth field, is what maintains the convective currents in the outer regions of the core? A few different methods have been suggested, but one that has the backing of several geophysicists is that due to the growth of the solid core. The solid core grows as the denser material of the outer settles down toward the inner core. As the denser material moves downward and the less dense material moves upward, there will be a change in the gravitational energy of the inner core, and this energy will be available to maintain the convective currents needed to generate the Earth's magnetic field.

Dynamo mechanisms have also been invoked to explain stellar magnetism. This proposal is clearest with respect to the behavior of sun-like stars. There seems to be a clear connection between stellar magnetic activity, rotation rate, convection, and age. At least some of the general principles that have been established can be explained in simple qualitative terms by a model proposed by Professor E.N. Parker of the University of Chicago. He suggests that a regenerative magnetic dynamo in a convective star that is rotating will produce magnetic fields that will inevitably rise to the surface. Turbulent convective surface motions will continually jostle these fields, and it is the jostling of the emerging fields that lead to plasma heating, and, consequently, to a chromosphere and a corona.

Chapter Three

Magnetic Mapping of the Universe—Galaxies

Section One—Our Milky Way Galaxy

Galaxies, in general, are vast collections of stars. Our own sun is one of about 100,000,000,000 stars that make up the Milky Way galaxy. However, the whole universe is made up of many thousands of galaxies that come in a variety of different sizes and forms. Most galaxies are not alone, but come in large clusters. Our own Milky Way is one of more than 20 galaxies that make up our local group. In this chapter we will first look at the magnetic fields of the Milky Way galaxy, and then at the magnetic fields of other galaxies. We will start the chapter by considering the size and structure of our own galaxy.

The Structure and Dynamics of the Milky Way Galaxy

In many cultures, the Milky Way is seen as a highway or road. For instance, the American Indian people, the Iroquois, see it as the "road for the souls" to the eternal kingdom; the Hindu religion says it is the path to the celestial throne of Aryaman; the Bushmen of the Kalahari Desert in South Africa think it is the reflection in the sky of the dying embers of campfires spread across the desert.

The Greek philosopher Aristotle thought the universe was divided into two separate regions, one below and one above the sphere of the moon. Above the sphere of the moon, the supra-lunar sphere, all was perfect and unchanging. In this region bodies were perfectly spherical and moved in perfect circles, or circles upon circles. In the sub-lunar sphere, however, things tended to be imperfect, always subject to change, and bodies moved in straight lines. Aristotle argued that, as the Milky Way was irregular, it could only belong to the sub-lunar region, and must therefore be an atmospheric phenomenon, so he included the study of it in his work on meteorology.

Scientific study of the Milky Way really started with Galileo, an Italian astronomer and physicist who lived from 1564 to 1642. In 1609 Galileo heard that someone in Holland had invented a device consisting of lenses that magnified distant objects. Within a few months he had made his own telescope, and with it he discovered that the Milky Way really consisted of a large number of stars.

William Herschel made the next great step in our understanding of our galaxy. He started life as a military musician in Germany, but he did not like the life, so he deserted and came to England, where he initially taught music. However, he had a great interest in astronomy, and taught himself to build telescopes. Using one of his large telescopes he counted the number of stars in different directions and different brightness in various parts of the sky, and used this to work out the shape of the Milky Way. This was done by assuming that all the stars were equally bright, and their apparent differences in brightness were solely the result of stars' varying distances from Earth (see Figure 3.1). Since Herschel's time the general size, structure, and dynamics of our galaxy has slowly been worked out, first with optical telescopes, and more recently with radio telescopes. Let us now look at the present view of the Milky Way, and the observations on which it is based.

A cross-section of the galaxy, at right angles to its plane, shows it to be a collection of about 100,000,000,000 stars spread out roughly in the form of a disc with a bulge, called the nucleus, in the center. Seen from above, a distinct spiral pattern would be evident for the distribution of the brighter stars (see Figure 3.2). All the stars are moving about the central bulge. Those

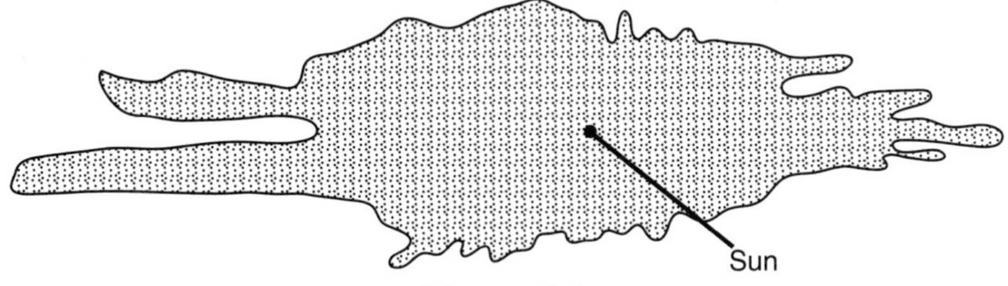

Figure 3.1
Herschel's view of the Milky Way.

near the plane of the Milky Way are moving nearly in circles, in ways that resemble, in certain respects, the movements of the planets around the sun. The stars are also moving with different speeds; those close to the center are moving faster than those further out.

Surrounding the whole galaxy is a halo of star clusters, with fairly large distances between neighboring clusters. The orbits of these clusters around the nucleus are more elliptical than those in the plane of the Milky Way (see Figure 3.3). Two of the most important problems that had to be solved in working out the structure of our own galaxy were finding the distances to the stars, and deducing the size of the galaxy as a whole.

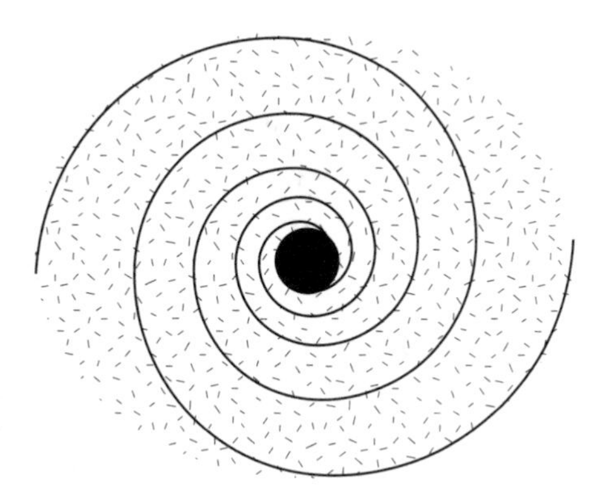

Figure 3.2
A diagram of the Milky Way as seen from above.

There are several ways of finding the distances to stars. For stars that are closer than 150 light-years, we can use the method of stellar parallax. This method is similar to that used by surveyors for measuring distances on the surface of Earth: If a surveyor wants to measure the width of a river, he or she can do so without having to cross it. The surveyor marks out a base line, then measures the two angles that an object—say, a tree—on the opposite bank make with each end of the base line (see Figure 3.4). By drawing to scale the triangle formed by the base line and the lines to the

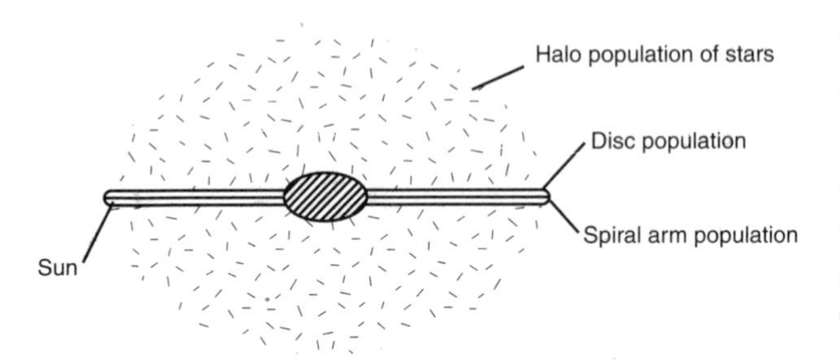

Figure 3.3
A side view of the Milky Way, showing the disc, the central bulge, and the halo.

tree, he or she can measure the distance to the tree on the drawing, and then scale this up to find the width of the river. The size of the Earth is too small in comparison to the distances to the stars to use this method in just this form. However, the apparent angular movement of a nearby star against the more distant stars, as observed from opposite points of the Earth's orbit, can be measured (see Figure 3.5). The diameter of the Earth's orbit (about 186,000,000 miles), rather than the diameter of the Earth (about 8,000 miles), is the effective base line in this case. This method becomes less accurate for stars more than 150 light-years from us (although this has been extended by the use of special Earth-orbiting satellites), because the angular movement becomes so small. However, as astronomy progressed further, it was soon discovered that other properties of stars could be used to find the distances to them.

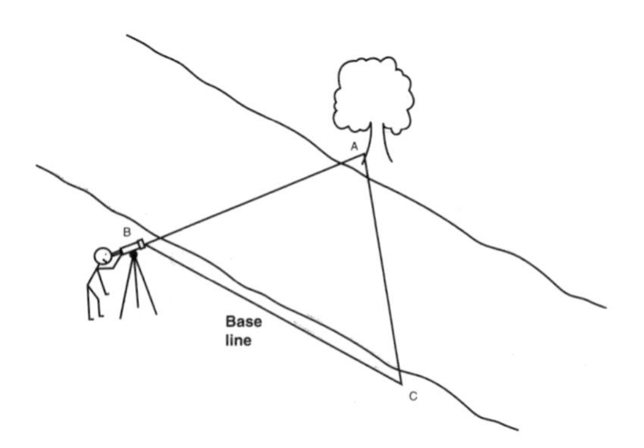

Figure 3.4
A surveyor measuring the width of a river.

The light we receive from a star depends not only on its intrinsic brightness, but also on its distance from the Earth. If we know how much light a star is emitting, and use a special light meter to measure the light we receive, then, by comparing the two brightnesses, we can work out its distance from the Earth. The number of watts marked on a light bulb is a measure of how much light it will emit, but how much we receive from it depends on how far we are from it (see Figure 3.6). Do stars carry labels telling us how bright they are? Yes, they do.

In the last chapter we saw how stars are classified using their spectra. We also saw that the spectrum for a given star was related to its temperature, and for all stars this was related to its actual brightness. The spectrum of a star is therefore a label that gives

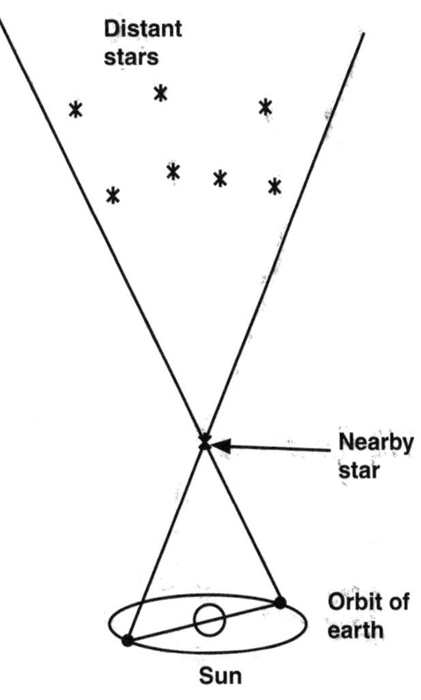

Figure 3.5

Nearby stars show parallax with respect to more distant stars as a result of the Earth's movement around the sun.

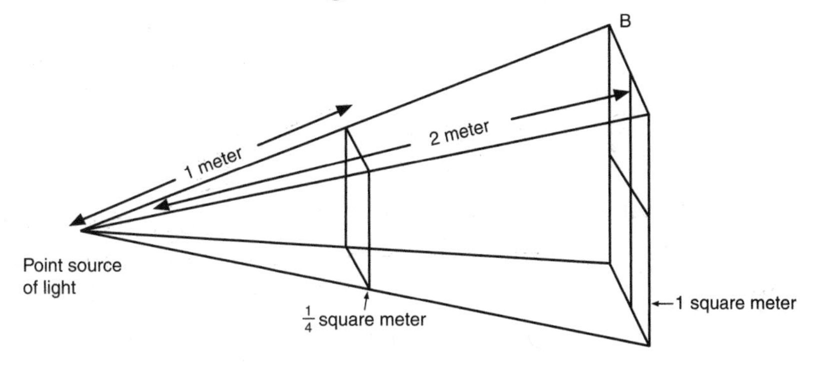

Figure 3.6

The inverse square law. The light passing through 1/4 m^2 at A passes through 1 m^2 at B. As a result, the light energy per unit area at B is one quarter of that at A.

us a good idea of how much light the star is emitting. The brightness of a star as seen from the surface of Earth can be measured, and if we compare this measurement with the actual brightness indicated by it spectrum, we can work out its distance from us.

Other labels carried by stars can give clues to actual brightness. One such is associated with a special type of star known as a *Cepheid variable*. The light emitted by a star of this type varies with periods ranging from a fraction of a day to several days. The period of variation is related to the average actual brightness of the star (see Figure 3.7). By comparing this with the brightness we measure on Earth, we can find its distance. This method works for Cepheid variables in our own and other galaxies.

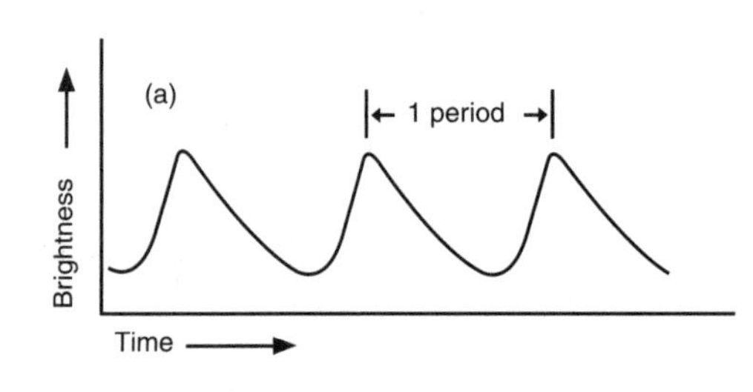

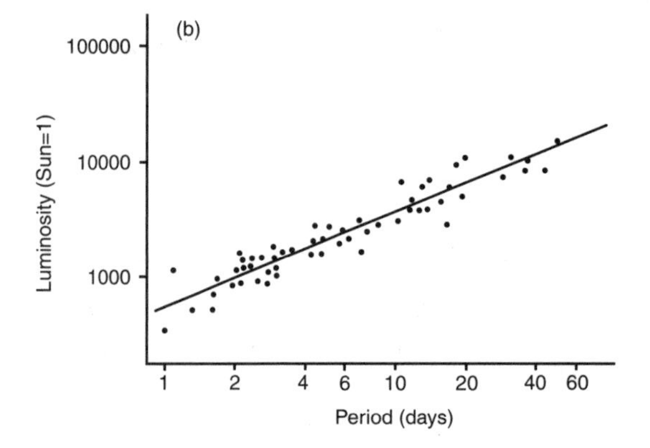

Figure 3.7
(a) Variations of the brightness with time of a Cepheid variable star.
(b) The relationship of brightness with period for Cepheid variable stars.

Although this method can give us some idea of the distances to such stars in our own Milky Way, we cannot use it to find the size of our galaxy, because there are tiny dust particles in the plane of the galaxy. These dust particles obscure the light from more distant parts of the Milky Way.

Earlier in this chapter we described the "halo" of the galaxy as a spherical distribution of stars that surround the whole Milky Way. The globular clusters of stars in this halo were used by the American astronomer Harlow Shapley (1885–1972) to estimate the size of the galaxy. He used three different methods to do this: First, some of the globular clusters contained

Cepheid variables, so he could use the method suitable for these stars to measure their distances. Second, he made some assumptions, one of which was that the brightest star in each cluster all had the same intrinsic brightness, and so by comparing the brightest star in a cluster of known distance with those in one of unknown distance, he was able to say how far away the latter cluster was. Third, he assumed that all clusters were of the same apparent size, and that observed differences in angular size, as seen from Earth, arose purely from differences in their distances from Earth. These last two methods may seem rather crude, but the three methods taken together gave Shapley a good estimate of the size of our galaxy. The methods used by Shapley worked because he was using clusters of stars that were out of the plane of the Galaxy, so the dust particles in the plane did not affect observations of the cluster stars.

Unlike the planets, all stars, even those near to us, are so far away that they only seem to change their relative positions in the night sky slightly, even throughout periods of several years, and even though they are moving at great speed. This means that astronomers have to use special methods to study the movements of the stars, as distinct from finding their distances from Earth. For some of the nearer stars, their movements can be measured by taking two sets of photographs several years apart. When the two sets are compared, small drifts in the positions of some of the stars become apparent. The distances to some of these stars are known, so we can calculate the speeds with which they are traveling across the direction in which we are looking (see Figure 3.8). However, this method does not tell us if the stars are moving toward or away from us.

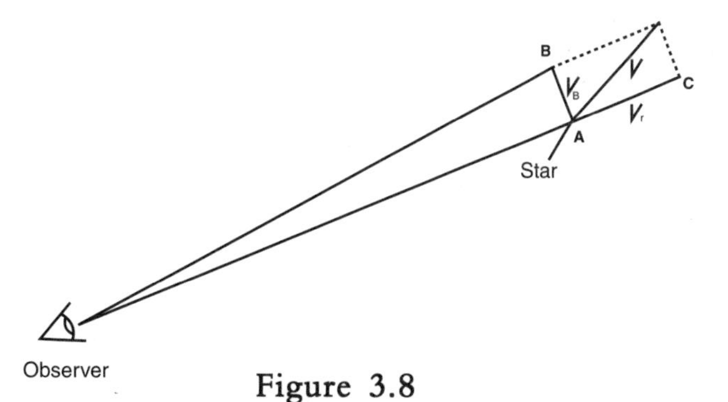

Figure 3.8

The star has a real velocity *V* in the direction shown. The observer can only measure directly the apparent change in position from A to B. The velocity *Vr* can be measured using the Doppler

To find out this additional piece of information, we must make use of the Doppler effect.

In the last chapter we saw that the dark absorption lines crossing the bright continuous spectrum of a star can be used to identify the elements present in its atmosphere. If the star is traveling toward us, then these dark lines will be shifted toward the blue end of the spectrum, and if they are travelling away from us they will be shifted toward the red end of the spectrum. The amount of shift toward either end tells us how fast the stars are traveling toward or away from us.

The two ways of measuring the movements of stars allowed astronomers to make important discoveries about the motions of stars in the galaxy. The motion of each star in a galaxy is controlled by the force of gravitation between it and all the other stars in the galaxy. Just as we can find the mass of the Earth by measuring the force of gravitation the Earth exerts on a particle, so we can measure the mass of the Milky Way (and other nearby galaxies) by measuring the force it exerts on stars—this force is deduced from the way they move. By studying the motions of stars at different distances from the center of a galaxy, we can also determine how the total mass of a galaxy is distributed within it.

Much of the early evidence for the existence of dark matter came from studies similar to these; astronomers studied the stars at the outer edges of our galaxy, and other galaxies, and found that these stars were moving in a way that indicated the presence of considerable amounts of matter beyond the visible limits of most galaxies. This matter did not emit light waves or radio waves, so its existence was only made manifest by the way it was tugging, gravitationally, on the outermost stars. Because it was invisible at all wavelengths, it was called *dark matter*. (We will discuss this matter in more detail in Chapter 5.)

Optical astronomy has given us a great deal of information on the size of our galaxy, the distances to stars, and their motions. Radio astronomy has considerably increased our knowledge about what is between the stars—the invisible part of the galaxy that does, nevertheless, manifest itself at radio wavelengths. Far from being a total vacuum, there is a great deal of material in interstellar space—atoms of gas (much of this gas is too cool to radiate light waves), subatomic particles, such as electrons (which move about and emit radio waves), and the dust particles first revealed by optical studies of distant stars. Radio waves can be used to detect different types of atoms or molecules, and to map the motions and distribution of these particles.

Hydrogen is the most abundant element in the interstellar medium, but several other atoms and molecules have been detected. Using the Doppler effect, radio astronomers have found the gas to be moving about the center of the galaxy in much the same way as the stars. Just as radio waves are not affected by fog on Earth, so radio waves from the galaxy are not affected by dust particles, which block light from the most distant parts of the galaxy. Radio telescopes can therefore study the whole galaxy, and our picture of the Milky Way is much more complete as a result of this work.

Cosmic Rays and the Galactic Magnetic Field

The first suggestion that the Milky Way galaxy may have a magnetic field was made by two American scientists, Subrahmanyan Chandrasekhar (1910–1995) and Enrico Fermi (1901–1954). They were trying to explain the existence of cosmic ray particles (very fast-moving subatomic particles), which were coming to Earth from all directions beyond the solar system. They suggested that these particles were ejected in violent supernova explosions. However, if this were the case, the particles would be expected to come largely from the plane of the Milky Way galaxy where supernova explosions occur. Instead, they came from all directions in space. Another problem posed by their suggestion concerned the total number of particles observed. If these particles were produced in supernova explosions, many would just be ejected from the galaxy, because their energies were too high for the gravitational tug of the whole galaxy to pull them back to the galactic plane. Both these difficulties could be resolved by supposing that a large-scale field existed in the plane of the galaxy. The cosmic-ray particles would then be confined to the plane of the galaxy, despite their high energies, by the interstellar magnetic field lines around which they would spiral. These field

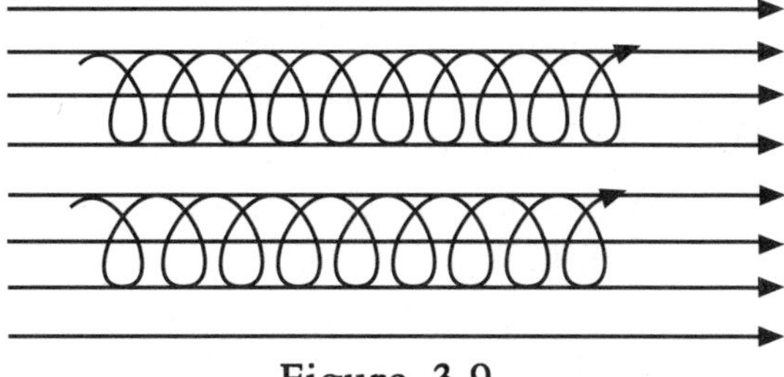

Figure 3.9

Containment of cosmic rays by the magnetic field of the Milky Way.

lines are frozen into the thermal gas of the interstellar medium, which is concentrated toward the plane of the galaxy by the gravitational field of the Milky Way as a whole. The cosmic ray particles would, because of the field lines, come to the Earth from all directions, because they would be spiraling around the lines rather than moving along them (see Figure 3.9). Much later it was pointed out, by E.N. Parker, that such a situation was unstable. The gas would tend to move to the plane of the galaxy where the force of gravitation is much stronger. The cosmic ray particles are hardly affected by gravity, and because of their energy would tend to move upward. The magnetic pressure between the lines of force, and the additional pressure from the cosmic ray particles, would tend to push the lines upward. However, because the field lines are embedded in the gas, the situation cannot simply be inverted. The gas can move along the field lines and the magnetic field can be distorted upward in between (see Figure 3.10).

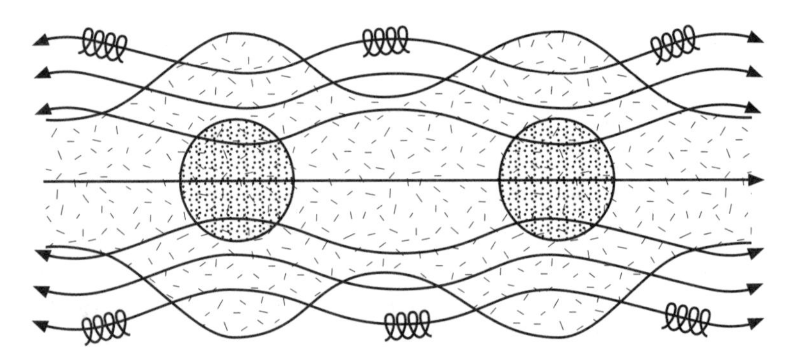

Figure 3.10
Instabilities between the interstellar gas, magnetic field, and cosmic ray particles.

Magnetic Fields and the Spiral Arms of the Galaxy

Chandrasekhar and Fermi also invoked the presence of a magnetic field to explain the spiral structure of the Milky Way. They assumed that the bright young stars of the galaxy, which were found in the spiral arms, were there because they were still close to their birth places in the interstellar gas, which was denser along the spiral arms. Yet if the gas was in the form of spiral tubes, there was not enough gas pressure to support such a tube against the gravitational force at right angles to the disc of the galaxy. However, if the galactic magnetic field was in the form of lines of force along the length of the arms, then there might be additional magnetic pressure at right angles to the lines of force, and this could supply the extra pressure needed to support the tubes of the spiral arms (see Figure 3.11).

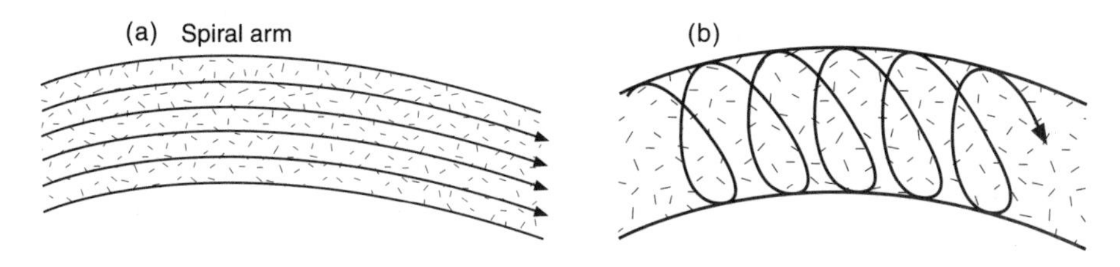

Figure 3.11
(a) Magnetic field lines directed along spiral arms.
(b) Magnetic field lines in the form of helices with axis along spiral arms.

Several years later, Fred Hoyle and John Ireland showed that such support could be provided by a field in the form of helical lines wound around the length of the tubes that formed the spiral arms. The modern theory of spiral structure is due to astronomers C.C. Lin and F.H. Shu. They suggested and developed a theory that, in effect, says that the spiral structure is due to density waves maintained by the self-gravitation of the large-scale distribution of matter in the disc. As a result of this work, magnetic fields are no longer considered important when discussing the spiral structure of the galaxy.

In 1948 Hall and Hiltner, two astronomers working at the U.S. Naval Observatory in Washington, built a polarimeter to measure the linear polarization of stars. Their reason for starting a program to measure stellar polarization was a theoretical prediction made by Chandrasekhar that, under certain circumstances, the light from eclipsing binary stars would be polarized. Chandrasekhar and Breen had suggested that the scattering of light by electrons was the chief source of opacity in early-type stars. (As we saw in the last chapter, stars are classified using letters, ranging from O and B to R and N. The name *early-type* was coined in the early 20th century, when it was erroneously believed that stars began at the O end and evolved toward the N end, so those at the O end were called *early-type stars*.) They

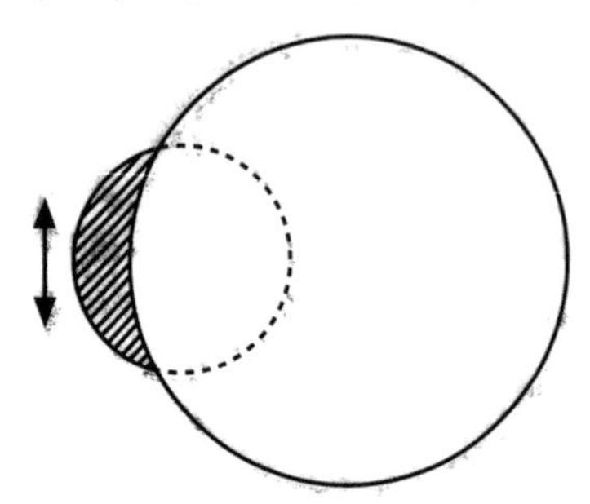

Figure 3.12 Polarization of light from certain binary systems.

suggested that if an early-type star was partially eclipsed by another star in a binary system, then it would be possible to detect that effect (see Figure 3.12).

Hall and Hiltner set out to do just that, but instead they found that a large number of stars, most of which were not binary, were polarized. Their studies showed that the polarization was not in any way related to spectral type. However, for stars in certain directions within the galaxy, the amount of polarization was related to the amount of reddening of the light from the star. This reddening was known to be the result of the absorption of blue light by dust grains in the interstellar medium, and it is not dissimilar from the reddening of light from the sun by dust grains in dry, desert-like conditions. This led to the theory that the polarization might be due to dust grains aligned by the galactic magnetic field.

Modern theories of interstellar polarization are based on the work of Leverett Davies, Jr., and Jesse L. Greenstein. It is now generally accepted that the polarization is due to scattering and absorption by plate-like dust particles. The magnetic properties of the particles are such that they spin around the lines of force, rather like a wheel on an axle. A beam of unpolarized light coming from a star will interact with these particles. The waves vibrating at right angles to the field lines will "see" more of the dust particles, and more of these waves will be absorbed. As a result of this interaction, the light passing through the interstellar dust will be slightly more strongly polarized parallel to the magnetic field lines (see Figure 3.13). This theory predicts that the polarization of starlight will be at maximum when looking at right angles to the field lines, and almost zero when looking along the field lines.

The polarization of a large number of stars has now been measured, and an analysis of the data seems to show that the magnetic field of the Milky Way is parallel to the plane of the galaxy. In the neighborhood of the sun, the observations are consistent with a field pointing in a direction making an angle of 50 degrees with the galactic center (see Figure 3.14).

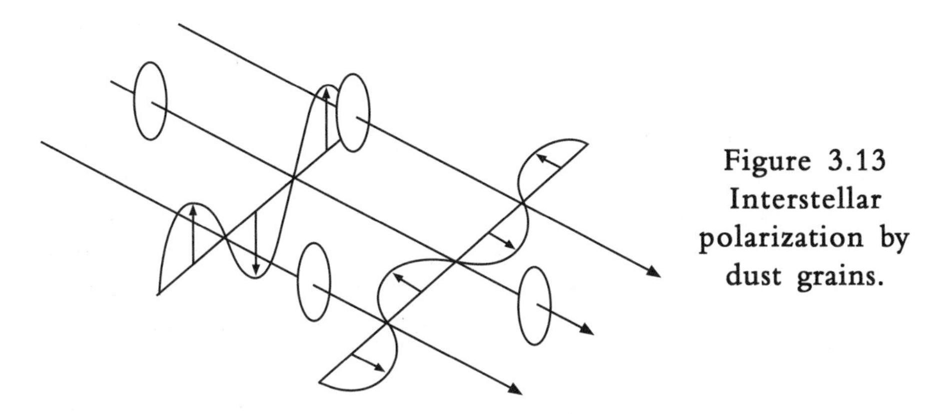

Figure 3.13 Interstellar polarization by dust grains.

Similar to all methods in astronomy, such that we do not have direct access to the objects which we are studying, this method of investigating the field has its advantages and disadvantages. An important advantage is that we can put distance limits on where the polarization is occurring, because it must be somewhere between Earth and the star we are measuring, and not beyond the star. (In 1966 I used this method to find the distance to a feature of the radio sky, called the *North Galactic Radio Spur*, which I showed to be due to an anomaly in the magnetic field of the galaxy.) One disadvantage is that the method can only be used to investigate the field in those regions where dust exists. This means that it is confined to the thin layer of dust near the plane of the Milky Way. Because the dust also reduces visibility in the galactic plane, the method is only applicable to the region in which the light from the stars has not been obscured by absorption.

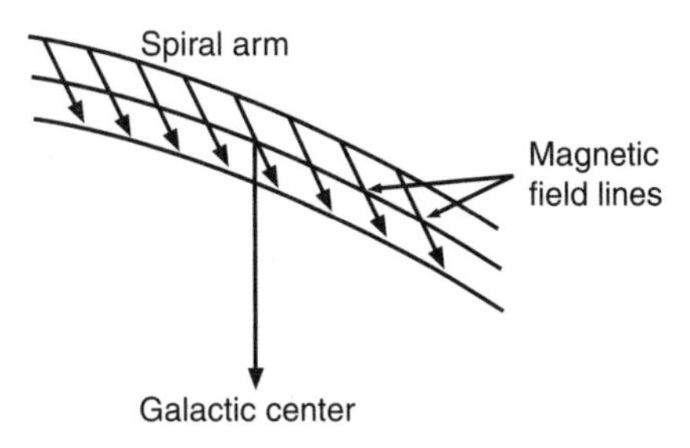

Figure 3.14 Magnetic field lines make an angle with the spiral arm axis.

Synchrotron Radiation From the Galaxy

The fact that not all radio radiation from the galaxy could be accounted for in terms of thermal emission from the gas was already evident soon after the first radio map of the Milky Way was published by Karl Jansky (1905–1950) in 1932. However, it took several more years before it was realized that part of the radio radiation was due to cosmic-ray electrons spiraling in the magnetic field of the Milky Way. In 1950, Hannes Alfven (1908–1995) and N. Herlofson proposed that the non-thermal part of the emission was produced by the synchrotron process when cosmic-ray electrons interacted with the trapping fields, or magnetospheres, of stars. This model was no doubt inspired by Alfven's picture of the solar system in which cosmic rays originated near the sun and were, to some extent, made isotropic by the interplanetary magnetic field. It was left to K.O. Kiepenheuer to make the suggestion that the radiation was due to electrons in the galactic magnetic field.

Mapping the structure of the galactic magnetic field using the polarization of synchrotron radiation is beset with difficulties: When a measurement is made in a given direction, we are in fact detecting all the radiation in the beam of the radio telescope. If, as it is now believed, the galactic magnetic field has small-scale fluctuations superimposed on a more ordered field, then the polarization from various parts of the interstellar medium within the beam will be oriented in different directions, and some of them will tend to cancel each other out. The observations are still further complicated by Faraday rotation, if the field being studied has a component along the line of sight. This is because the polarization of elements at different distances from the telescope will be rotated by differing amounts, and the net result will be a decrease in the percentage of polarization. However, despite these difficulties, it is possible to make deductions about the local field, where the effects just mentioned will be less important. When looking at right angles to the field lines, there will be no Faraday rotation, and the radio polarization should be at right angles to the optical polarization coming from this region. This has in fact been observed near the plane of the galaxy in a direction making an angle of 140 degrees with the galactic center. This method gives a direction of 50 degrees for the local magnetic field, which is consistent with that given by optical polarization.

Faraday Rotation Measures of Extragalactic Radio Sources

Radio-astronomical observations of Faraday rotation in the radiation from extragalactic radio sources have proved to be one of the most effective ways of investigating the large-scale structure of the galactic magnetic field. The possibility that Faraday rotation may be occurring somewhere in open space was first suggested by B.F.C. Cooper and R.M. Price in 1962, when they noticed that the angle of polarization from a radio source—called Centaurus A—varied with wavelength. As the number of extragalactic radio sources with measured rotation increased, some researchers used the data to argue that the field pointed in opposite directions above and below the plane of the galaxy. This type of conclusion is only possible with Faraday rotation measures, because the other methods discussed earlier can only give information on direction, but not on the sign of the field—they cannot be used to tell if the field is pointing toward or away from the observer.

After this early work there arose support from different quarters for three models of the magnetic field. The first one consisted of a loop of magnetic field superimposed on a longitudinal field directed along the local spiral arm. The second model has a longitudinal field directed along the spiral arm, but it points in one direction above the plane of the galaxy, and in the opposite direction below the plane of the galaxy, with a zero field in the actual plane. The third was a sheared helical field wrapped around the axis of the local spiral arm. (These models are illustrated in Figures 3.15, 3.16, and 3.17.)

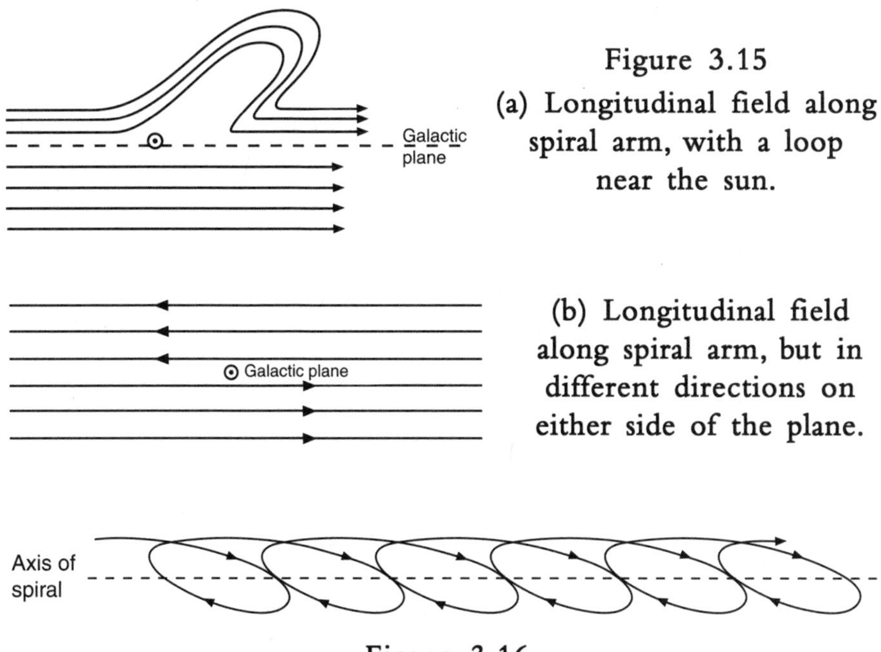

Figure 3.15
(a) Longitudinal field along spiral arm, with a loop near the sun.

(b) Longitudinal field along spiral arm, but in different directions on either side of the plane.

Figure 3.16
A sheared helical with axis along the spiral arm.

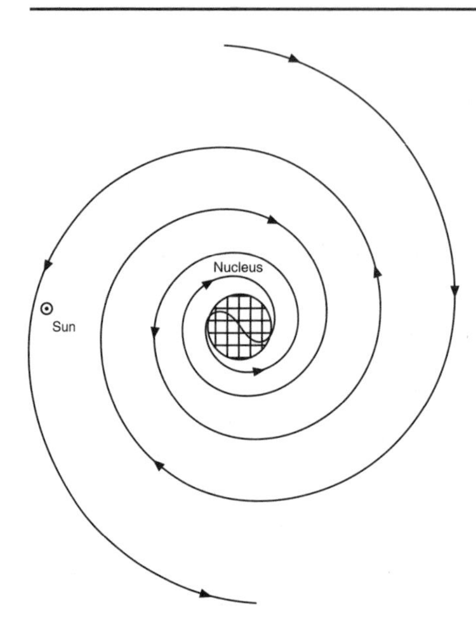

Figure 3.17 Bisymmetrical spiral arm field.

Pulsars and the Interstellar Medium

Pulsars have also been used to give information on the properties of the interstellar medium. Although all types of electromagnetic radiation travel with the same speed in free space, they do not travel with the same speed when neutral or ionized gases are present in space. The pulses from pulsars can be studied at different wavelengths by tuning radio telescopes to different frequencies. These observations show that the arrival times of the pulses at different wavelengths differ by small but measurable intervals. From these intervals it is possible to work out the different speeds of varying wavelengths in the interstellar medium. This gives some information on the electron density in space. Alternatively, if an average electron density is assumed, from other types of measurement, it is possible to make an estimate of the distances to the pulsars.

The radiation from pulsars is also linearly polarized, and the angle of polarisation varies with wavelength. This is due to the Faraday effect, and hence this effect can be used to map the galactic magnetic field in much the same way as rotation measures from extragalactic radio sources can be used. All these observations of rotation measures give us information on the field in those regions where thermal electrons are present.

Unified Models for the Large-Scale Galactic Magnetic Field

Model Based on Distorted Concentric Circles

In 1967 I presented my doctor of philosophy thesis at Manchester University, in which I proposed a unified theory for the galactic magnetic field. I also defended this model in my book *Cosmic Magnetism*, which was published in 1986. In that book I said, "Using information from all available sources, there is general support for a model in which the field lines are concentric circles, about the center, and in the galactic plane. However, these circles have been disturbed by shock waves associated with the formation of spiral arms. Such a field is consistent with a galactic dynamo...."A diagram illustrating this model is given in Figure 3.18.

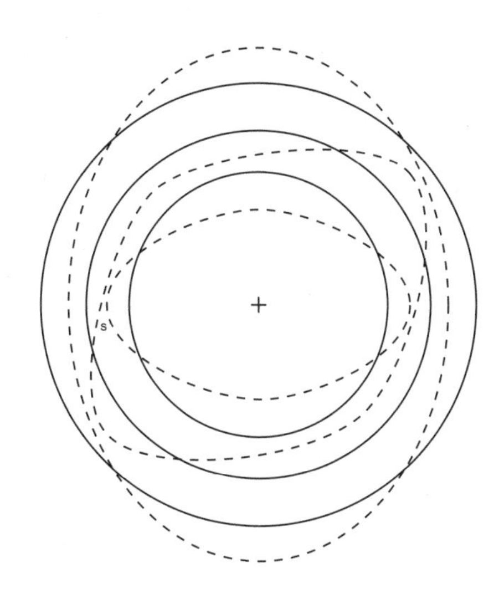

Figure 3.18
Author's model of the large-scale structure of the galactic magnetic field. The magnetic field lines (represented by the dotted curves) are in the form of concentric circles which have been distorted by the density wave pattern that generates the spiral patterns of bright young stars seen in many spiral galaxies. (From the author's PhD thesis, Manchester University, 1967.)

In *Cosmic Magnetism* I also said:

> Recently two astronomers in Canada, Simard-Normandin and Kronberg, proposed a bisymmetric spiral field model, in which the field lines followed the spiral arms but reversed when going from one spiral arm to the next. This particular model has been supported by two Japanese astronomers, Sofue and Fujimoto. These two argue that the field is consistent with the hypothesis of a primordial origin to the field. If the whole universe possessed a magnetic field, the field would be frozen into the gaseous material out of which the galaxies formed. The gravitational collapse of this material to form a galaxy would have led to an amplifying of the field and the winding-up of the field to follow the spiral arms would have resulted from differential rotation. However, the age of the Galaxy is such that, if this explanation were correct, there should be several more spiral arms than have been observed, and they should be more tightly bound.

At that stage I was still unconvinced by the evidence to support the bisymmetric spiral field model. Although I still feel there are problems with a primordial origin for the field, the evidence is now more convincing, for a bisymmetric model, than it was in 1986.

Writing in *Astrophysical Journal* in 2006, R.N. Manchester, A.G. Lyne, G.J. Qiao, and W. van Straten used a variety of observations in their attempts to deduce the structure of the galactic magnetic field, and they came to the following conclusion: "We propose that the large-scale galactic magnetic field has a bisymmetric structure with reversals on the boundaries of the spiral arms. Streaming motions associated with spiral density waves can directly generate such a structure from an initial inwardly directed field."

SECTION TWO—OTHER GALAXIES

In the last 25 years it has become clear that magnetic fields exist in many different types of galaxies, and that in certain classes of extragalactic objects, they play an extremely important role. In this section we review, briefly, what is known about the structure and dynamics of these distant objects, with particular emphasis on what radio and optical astronomy have together taught us about their magnetic fields.

The Realm of the Nebulae

Early in the 20th century arguments still rage about the nature of the nebulae. The application of spectroscopy to astronomical problems had shown that some nebulae were in gaseous components of our own Milky Way galaxy, but the nature of the spiral nebulae was still a mystery. The argument was resolved in 1924 by Edwin Hubble (1889–1953), using the 100-inch telescope at Mount Wilson. With this instrument he discovered Cepheid variables in the great Andromeda spiral nebula and in several other spirals. By using the period-luminosity relationship, discussed in the first section of this chapter, he was able to find the distances to these objects.

Using this method as a basis, Hubble employed other methods to find the distances to more distant galaxies for which the Cepheid variable method did not work. One such method made use of supergiant stars, which are much brighter than Cepheids, and so could be seen over larger distances. He assumed that the brightest supergiants in all galaxies have the same absolute brightness, and so, observed differences in brightness would be solely due to variations in distance. However, at greater distances even the supergiants could not be used as distance indicators.

At this stage Hubble started using the properties of galaxies themselves. He found that many galaxies—though not all—tend to occur in clusters; some clusters contain only a few galaxies, whereas others contain many hundreds. He then measured the distances to some of these clusters by assuming that the brightest galaxies in a cluster had the same intrinsic brightness.

The Hubble distance scale had to be revised in 1952, when Walter Baade (1893–1960) discovered that there are two types of Cepheid variables, and that they have different period-luminosity relations. All the Cepheids in the disc of our own Milky Way galaxy are of Type II, so distances in the plane of our own galaxy using these stars did not have to be revised. However, the Cepheids in the globular clusters of the halo of our own galaxy and in other galaxies were of Type I, and this meant that all distances measured using the period-luminosity relation for Type II had to be increased by a factor of 5.

For some intermediate distant galaxies it is possible to use novae as distance indicators. A nova is a star that suddenly grows extremely bright, sometimes throughout a period of several hours, and then fades away throughout a period of days. The time taken for a nova to fade to a particular fraction of its maximum brightness is related to the actual value of this maximum. Novae are occasionally seen in our own galaxy, and in that case it is possible to measure their distances. From this it is possible to calculate a nova's absolute brightness, and then relate this to the rate at which it fades away. This relationship can then be used to find the distances to novae in other galaxies.

Hubble combined his distance measurements for galaxies with Doppler measurements of galaxy velocities made by Vesto Melvin Slipher, at the Lowell Observatory between 1912 and 1924, to deduce Hubble's law on the recession of galaxies. This law states that the speeds with which galaxies are moving away from us is directly proportional to their distances from us. This means that the most distant galaxies are traveling away from our galaxy faster than the nearby ones. Most astronomers now believe Hubble's law is a direct consequence of the big bang theory for the origin of the universe. According to this theory, all matter in the universe was originally concentrated at very high densities in an extremely small region of space, and this high concentration of matter produced an explosion that sent fragments of matter shooting out in all directions. All the galaxies and stars eventually formed out of these fragments of matter. Fragments given the highest speeds by the initial explosion were pushed out farthest, and one would expect these objects farther away from the center of the explosion to be moving faster than those nearer the centre. An important consequence of this type of expansion is that, from the point of view of any galaxy in the universe, every other galaxy will be moving away with a speed directly proportional to its distance from the observer's galaxy.

Hubble's law can be used to find the distances to galaxies for which other methods are inadequate: There are absorption or emission spectral lines in the spectra of most galaxies, and by comparing measurements of these lines with spectral lines generated in the laboratory, it is possible, using the Doppler effect, to calculate the speed with which galaxies are moving away from us. Once this has been done, it is possible to estimate the distances to these galaxies using Hubble's law (see Figure 3.19). This method has proved to be useful for radio astronomers because they have no other direct way of measuring the distances to extragalactic radio sources.

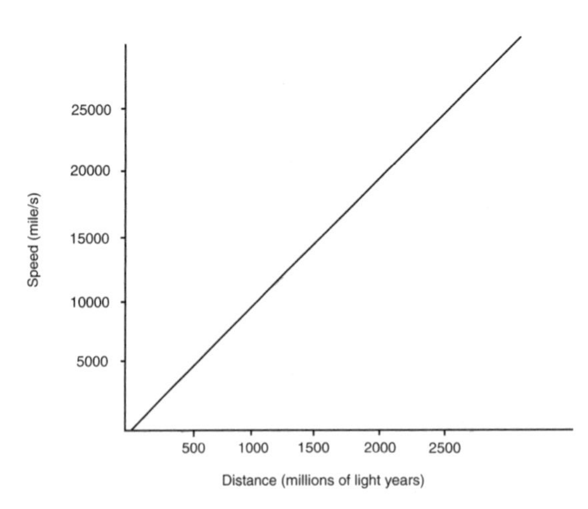

Figure 3.19
Graph illustrating Hubble's law.

The system used by optical astronomers to classify galaxies also started with Edwin Hubble. Normal spiral galaxies are similar to our own, in that they have two spiral arms surrounding a more compact central nucleus. How tightly the spiral arms are wound up varies from one galaxy to another, and this leads to further subdivisions within the class. Barred spiral galaxies differ from normal spirals in that they have a bar across the center rather than an ellipsoidal nucleus. They also have two spiral arms, but these are not as tightly wound as in a normal spiral. About 60 percent of galaxies are normal spirals, and about 20 percent are barred spirals. The rest are a mixture of other types. Elliptical galaxies are, as their name implies, elliptical in shape. They vary from being almost spherical to being highly elliptical, and there are other differences within the class. Irregular galaxies are rather shapeless. The two best known examples of this type are the large and small Magellanic clouds, which can be seen quite easily in the night sky of the southern hemisphere.

Early in the history of radio astronomy, small, intense sources of radio radiation were found at fixed points in the sky. By about 1950, the positions of some of these sources were known with enough precision for optical astronomers to find them with giant telescopes. Some of the sources were identified with objects in our own Milky Way galaxy, while others were identified with irregular galaxies that had unusual properties in their spectra. Radio astronomers soon discovered that the radiation from most of these sources consisted of two or more areas of strong emission, and that the radiation was highly polarized (see Figure 3.20). These sources are now called *radio galaxies.* The way the intensity of radiation from these sources varies with the frequency to which the radio telescope is tuned suggests that it arises from the synchrotron process. This is confirmed by the polarization of the radio galaxies.

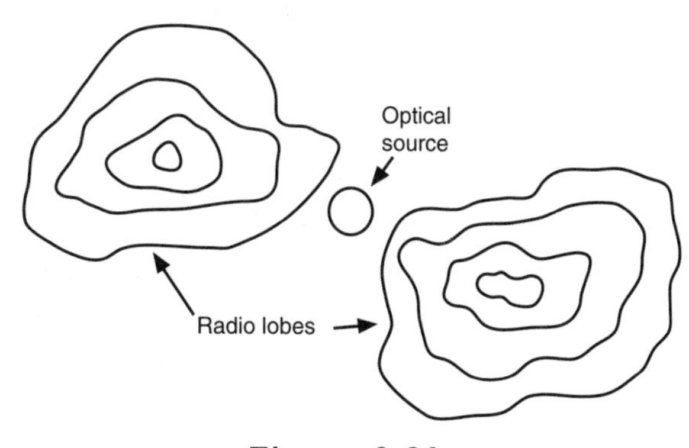

Figure 3.20
The structure of a typical radio galaxy. The curves represent contours of radio intensity.

Quasars

When Hubble's law was applied to the speeds of some of these radio sources (which were obtained by applying the Doppler effect to their optical spectra), it showed that they were more distant than most galaxies.

Some of these sources were in exactly the same part of the sky as optical sources that were so small they looked to be stars. However, their spectra could only be understood if some of the emission lines were assumed to be shifted very much to the red end of the spectrum. If Hubble's law was applied to these redshifts, it meant they must be at great distances from us, yet they were among the most powerful sources in the radio sky. This suggested that they must be extremely compact and able to produce a great deal of energy in a small volume. Radio astronomers called these sources *quasi-stellar radio sources*—or *quasars*.

Their compact nature has been confirmed by two separate sets of observations: The brightness of quasars varies on a timescale measured in months or weeks. This implies that the quasars are a few light-months or light-weeks across. If this were not the case (if they were not compact), light from different parts of the quasar would arrive at the observer at different times, and would thus mask the variations. The second set of observations concerns more precise measurements of angular size. The amount of detail that can be seen with a telescope depends on the wavelength being used and the size of the telescope. Because radio wavelengths are much longer than optical wavelengths, radio telescopes have to be much bigger than optical ones. The sizes of single-dish telescopes are limited by engineering considerations, but this limitation can be overcome by using a few separate telescopes, and then combining the information collected from a certain object at a particular time by each dish. Such a device is called an *interferometer*. The largest such devices available on Earth consists of separate dishes in different continents. In these large interferometers, the signals from the source being studied are recorded separately by each dish, together with precise information about times, and then the signals are later combined using a computer. Interferometers have been used to confirm that quasars have very small angular sizes. Because they are also at great distances from Earth, their actual sizes must be very small indeed.

Magnetic Fields in Other Galaxies

Optical polarization measurements have been carried out on several galaxies, including the Magellanic clouds, the Andromeda galaxy, and the Sombrero galaxy. As with the Milky Way, the polarization of light in these galaxies is believed to be due to the alignment of dust particles in the magnetic fields.

A typical set of results is the Sombrero galaxy. This galaxy is at a distance of 40 million light years from ours. It is seen almost edge-on, and a prominent feature of the object is the dark lane of dust across its equatorial plane. Above and below this dust lane is a lens-shaped luminosity that arises from an oblate distribution of stars. The polarization map shows fairly regular polarization close to the lane and parallel to its length. This would imply that the magnetic field of the galaxy is parallel to its central plane. Just above the nucleus, the polarization is at right angles to the lane. W.S. Pallister, S.M. Scarrott, and R.G. Bingham, who made the observation, suggest that this is due to reflection of light from the dust grains, rather than the transmission of light through the grains, which is the case for most of the galaxy.

Optical polarization measurements of spiral galaxies have been supplemented by radio astronomical measurements; by analogy with our own Milky Way, it is believed that the radio polarization results from the synchrotron process at work in these objects. Faraday rotation and the degree and the direction of polarization (after correction for Faraday rotation) can be used to discover properties of the field strength and structure. The information obtainable from such studies is illustrated by discussing the Andromeda galaxy.

R. Beck, a radio astronomer working at the Max Planck Institute of Radio-Astronomy in Germany, made a detailed study of Andromeda. By studying the radio polarization in this galaxy at 11.1 cm, he was able to deduce that this object had a large-scale magnetic field. This field is aligned along the spiral arms of neutral hydrogen. It forms an elliptical ring-like structure in the plane of the galaxy at about 30,000 light-years from the center. His observations showed no large-scale reversals of the magnetic field in the region between 7 and 16 kiloparsecs (1 kpc = 3,260 light-years) on a scale equal to the interarm spacing of 2–3 kpc.

A Canadian astronomer, Jacques Vallee, has used observations on the magnetic fields in spiral galaxies to reach some general conclusions about their structure. He divided these galaxies into two separate classes: One class was composed of galaxies with stellar spiral arms and spiral magnetic fields distributed over most of the outer arms, and the other class consisted of galaxies with spiral arms and a circular type of magnetic field over large parts of the outer spiral arms. He showed that the first class all had nearby companion galaxies, and suggested that the tidal effects from such a companion could possibly create material arms and align stars, gas, and magnetic fields in spiral arms. The second class did not have nearby companions, and, consequently, had much smaller tidal effects from other galaxies. He also suggested that these smaller tidal effects could only excite density wave arms, with the stars forming a spiral structure, but the gas and magnetic fields forming nearly circular structures.

The polarization at radio wavelengths of several radio galaxies has also been studied, and a much higher degree of polarization is shown in these galaxies than in normal spiral galaxies. The strength of the fields in these objects is much higher than in normal galaxies, and the fields play a much larger part in the structure and evolution of these radio galaxies than they do in other galaxies. Some scientists now believe that the highest-energy cosmic ray particles that reach the Earth were probably generated within radio galaxies, and that their magnetic fields play an important role in accelerating these particles to high energies.

Throughout the last 30 years a theory has been proposed that explains most of the properties of radio galaxies and quasars. The suggestion is that at the center of any one of these objects there exists a massive black hole. The black hole will, because of its very high gravitation field, attract matter toward it, but it will also swallow up all the light close to its ever-shrinking surface. Suppose that the black hole is rotating, and around it is an accretion disc of matter containing a magnetic field (see Figure 3.21). This field extends for thousands of light-years into space. If the magnetic field were rotating with the black hole, it would mean that the outer parts of the field would be moving faster than light. Einstein's special theory of relativity tells us that this cannot happen, and therefore it

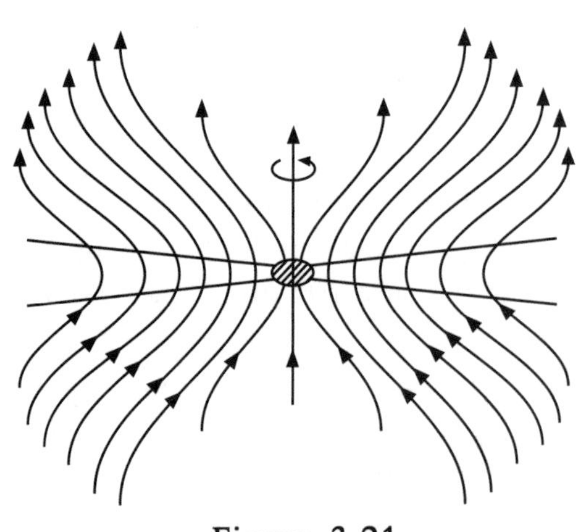

Figure 3.21
Accretion disc around a rotating black hole with a magnetic field.

has been suggested that the magnetic field "slips" with respect to the surface of the black hole. The slipping of the field would generate strong electric fields in the neighborhood of the hole, and thus a giant dynamo would be created. According to modern theories of particle physics, electrons and positrons (positively charged electrons) will be created out of the energy of the field. The electric field then propels these particles along two beams, in which they are guided by the magnetic field (Figure 3.22). These beams of particles then give rise to two regions of intense radio emission on either side of a much smaller visible source.

Are There Magnetic Fields in Intergalactic Space?

In 1968 three Japanese astronomers, Y. Sofue, M. Fujimoto, and K. Kawabata, claimed they had found evidence for a contribution from an intergalactic medium to the observed rotation measures integrated over elliptical galaxies and quasars. Their suggestion, together with later work by other astronomers, was based on the large-scale asymmetry in the Faraday rotation measures of extragalactic radio sources, and its possible correlation with the redshifts of these sources. They claimed that this could be explained by a weak homogeneous magnetic field stretching over cosmological distances. Jacques Vallee, from Canada, has used a much more extensive set of data on Faraday rotation measures to show that this claim was not confirmed by the larger sample of data, and that if such a field did exist, it must be very weak indeed. However, if such a field did exist, even if it is weak, it must have been stronger in the past, and this being so, it could have consequences

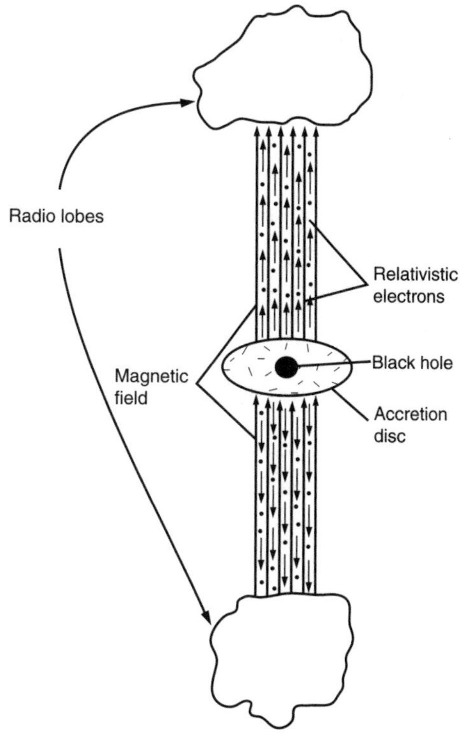

Figure 3.22
Model of a radio galaxy.

for cosmology and the formation of galaxies. It is therefore worth expanding briefly on the possibility of intergalactic magnetic fields and their possible consequences.

The presence of a weak homogeneous field in the universe at the moment implies, because of the cosmological expansion of the universe, a much stronger homogeneous field in the early universe. Such a homogeneous field would have allowed expansion to take place faster in the direction of the field than at right angles to the lines of force. This nonuniform expansion would have observable consequences, in that the remnant heat radiation from the early universe would be different in different directions. The high degree of uniformity observed from this background radiation places further limits on the strength of the magnetic field that may have existed in the early universe.

The presence of the magnetic field in the interstellar medium can influence the growth of instabilities in the clouds out of which stars are formed, in much the same way that a magnetic field in the early universe could influence the development of instabilities that lead to the formation of galaxies. Once galaxies are formed, the magnetic field could affect the spinning of the galaxy and how the spin rate varies with distance from the center. Detailed investigation of these possible effects requires more information on the present strength and uniformity of the field.

Because magnetic fields are not intrinsic properties of space or matter, the existence of a primordial magnetic field does raise questions concerning its origin. Some theorists have proposed a mechanism by means of which a magnetic field could be generated in an expanding plasma dominated by radiation. These are the kinds of conditions that prevailed in the early universe, and can best be described by looking at what happens to a uniformly rotating and expanding whirlpool of matter and radiation, called a vortex. While the vortex is expanding, the total mass of matter, the total density of radiation and the angular momentum of both, are separately conserved. In the plasma, protons collide with the much heavier neutral atoms, and, as a result, the two components of the plasma move with the same angular speed. The protons do not, however, interact with the radiation to the same extent as the electrons, and, consequently, the radiation is more effective in braking the rotation of the electrons. This means that the protons and the electrons move at different speeds, and this gives rise to an electric current that will generate a magnetic field. The application of these principles to the early universe shows that it is possible for a primordial magnetic field to have been generated.

The matter of intergalactic magnetic fields has recently been discussed (April 2005) by two astrophysicists, Arno Dar, from Israel, and A. de Rujula, from Switzerland. In a paper called "Magnetic Fields in Galaxies, Galaxy Clusters, and the Intergalactic Space," they say: "We show that, if the turbulent motions induced by the winds [from supernovae explosions] and the cosmic rays generated magnetic fields in rough energy equipartition, the predicted field strengths coincide with the ones observed not only in galaxies...but also in clusters...." From their calculations, they also predicted that the strength of the intergalactic magnetic field would be about 1,000 times weaker than the interstellar fields.

Chapter Four
The Sun and Its Magnetism

This chapter looks at the nature and behavior of the magnetic fields of the sun and their extension into interplanetary space. Because these fields are intimately bound up with the structure and mechanics of the sun, the dynamics of its photosphere, chromosphere, and the corona, it is necessary to start with a brief review of some relevant facts about our sun.

Section One—The Basic Observational Details

The Sun

The sun is a fairly ordinary star. It is, however, the star about which the planets of the solar system move, and it is the ultimate source of heat, light, and energy for life on Earth. Because it is the most dominant extraterrestrial object, it is not surprising that many peoples in different parts of the world should have worshiped it as a god, and that it has been a major part of the serious study of astronomy for thousands of years. The points on the horizon at which the sun rises and sets at different times of year were the basis for the first social use of astronomy—that of calendar-making—for many ancient cultures. The varying lengths of the shadows cast by the sun were one of the earliest methods of time-keeping. The sun was also used by Eratosthenes—the keeper of the library at Alexandria—to measure the size of the Earth in 240 BC. In more recent times, throughout the last 400 to 500 years, the sun has proved to be an invaluable aid to navigation.

The study of the detailed physical nature of the sun had to await the invention of the telescope and the development of spectroscopy. These developments revealed the detailed behavior of sunspots and the chemical composition of the sun's atmosphere. Progress in atomic and nuclear physics led to the construction of mathematical models of solar and stellar interiors.

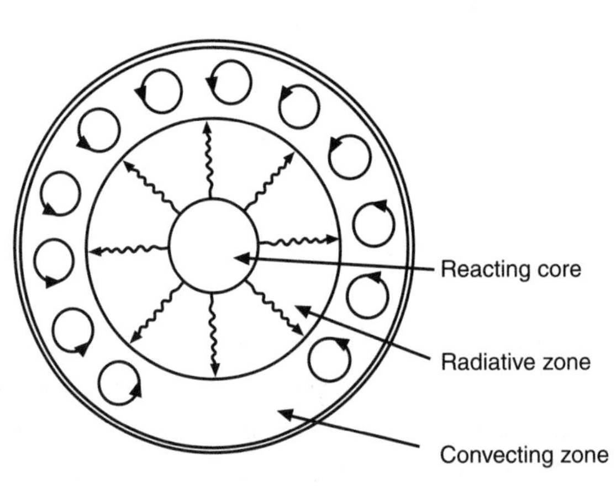

Figure 4.1
Internal structure of the sun.

According to the currently accepted model of the sun, its interior consists of three main regions: the core, the radiative zone, and the convective zone (see Figure 4.1). The core energy is generated by nuclear reactions, largely the conversion of hydrogen into helium. This energy is transported outward from the core by radiation in the radiative zone, and by convection in the convective zone. The magnetic field of the sun is believed to be due to a dynamo action that exists in the convective zone. Here the thermal convective currents of ionized plasma amplify any weak remnant field, due to the collapse of the sun out of the gases of the interstellar medium, to create a stronger field. As we will see later in this chapter, this field plays a vital role in solar activity.

The visible surface of the sun is called the *photosphere*, and it is the source of the characteristic absorption spectrum of the sun. Where the photosphere has a boundary with the convective zone, its temperature is about

6,000 degrees Kelvin. The photospheric temperature drops to 4,000 K where it borders with the more tenuous solar atmosphere. The temperature of the chromosphere then increases outward and reaches 50,000 K, where it meets the outermost part of the solar atmosphere, the corona.

Sunspots

Sunspots were first seen with the aid of telescopes in about 1610, and ever since then they have been of intense interest to astronomers. Although they appear to be very dark regions of the photosphere, this is entirely a contrast effect. The region of the sunspot is cooler than the surrounding areas of the Sun, so it emits less light and, as a result, it looks darker. However, most sunspots are as bright as the full moon. When a sunspot approaches the limb of the sun, the near side becomes practically invisible, whereas the far side is enlarged—this is called the *Wilson effect*, and indicates that sunspots are depressions in the photosphere (Figure 4.2). More often than not, sunspots occur in pairs or more complex groups, and whereas the smaller spots, known as pores, last only a few hours, the larger ones last for several days. Observations on the longer-lasting spots show that the sun is rotating, but not as a solid body. The rotation rate near the poles is about 37 days, whereas it is 26 days in the equatorial region.

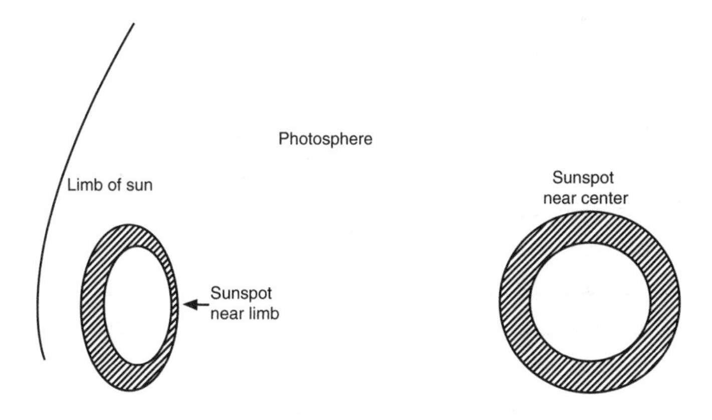

Figure 4.2
The Wilson effect.

The number of spots varies and reaches a maximum roughly every 11 years, but the period between successive maxima can be as short as seven or as long as 17 years. Each cycle begins with the formation of spots in middle latitudes of each hemisphere, about 40 to 50 degrees from the equator. Subsequently, spots form in lower latitudes until most of the photosphere is covered. After maximum coverage, it is the high-latitude spots that disappear first, and the last to fade completely are those close to the sun's equator. This behavior is very graphically shown in the Maunder Butterfly diagram, which shows the location of sunspot formation versus time in years.

Other Signs of Solar Activity

Photosphere faculae are bright areas that eventually engulf a sunspot group, but are usually noticeable before the appearance of the sunspots, and often last longer than the spots themselves. They are the first signs of solar activity, and when they are near the limb of the sun they appear brighter than the photosphere. Associated with the photospheric faculae are chromospheric faculae, or plages, which occur in the chromosphere just above the photosphere. These, however, can only be seen in photographs of the sun taken with special filters.

A flare is a sudden, local increase in the surface brightness of the sun, occurring in a region active with faculae and sunspots. The effect is produced by the sudden release of tremendous amounts of energy in the upper chromosphere, and is the culmination of activity that has been building up in the sunspot region.

Solar prominences are another visible consequence of solar activity. When seen at the limb of the sun, they appear as luminous arch-like structures with continual internal motion, but when projected on the luminous disc of the sun they just appear to be dark filaments. Sunspot prominences (also sometimes called *active prominences*) appear over a sunspot group, whereas quiescent prominences are associated with particular regions without sunspots or with decaying groups of spots.

In 1960 an astronomer called Leighton used the Doppler effect to investigate motions in the sun's atmosphere. He discovered that the sun seems to be divided into about 1,000 large cells by the horizontal velocity of the photospheric gas, which changes in speed and direction from one cell to the next. These regions are known as *supergranules* to distinguish them from

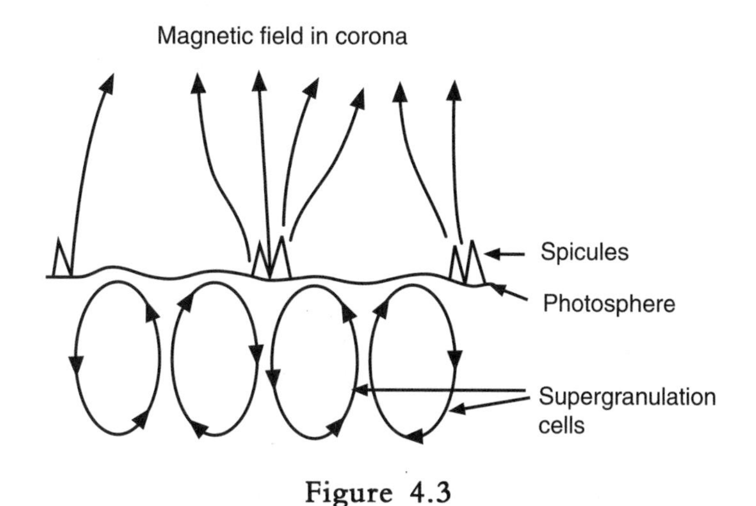

Figure 4.3
Spicules, supergranulation, and the solar magnetic field.

the well-known granular pattern. The ordinary granular pattern is due to the convective motions in the convective zone of the sun. At the boundaries of these supergranules there is a concentration of many small jets, known as *spicules*, which shoot out through the chromosphere into the corona (see Figure 4.3).

Having reviewed the general features of solar activity, we can now look at observations of the magnetic fields of the Sun and relate some of these to the dynamics of the features just described.

Solar Magnetic Fields

The Zeeman effect (described in Chapter 1) was first applied to astronomical objects by George Hale at Mount Wilson Observatory, when he used it to measure the magnetic field strengths of sunspots. He was able to show that when sunspots occurred in pairs, the magnetic field emerged from below the sun's surface at one spot, and entered the surface again at the other spot. Moreover, during one 11-year sunspot maximum, the spots in the two hemispheres had different east-west polarities from each other, but during the next cycle the polarities in each hemisphere would be reversed (see Figure 4.4).

Later work showed that the sun has fields of opposite polarity at its two poles, and that this polarity changes at the start of a new sunspot cycle.

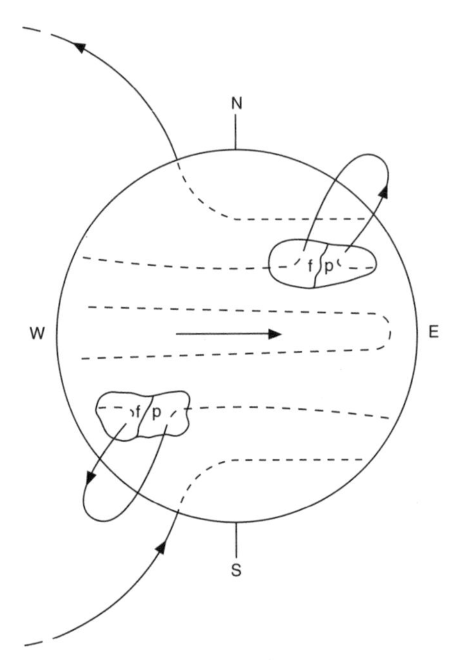

Figure 4.4
The sunspot polarity in the north hemisphere is opposite from that in the south hemisphere.

There is also a weak large-scale field, which, similar to the spicules, seems to concentrate at the boundaries of the supergranular cells, in quiet areas of the sun. Magnetic fields are also found in prominences seen against the disc, and quite often these occur between the magnetic areas of opposite polarity associated with sunspot pairs. Astronomers have used special mathematical techniques to analyze the data on the magnetic field distribution over the entire surface of the sun, and from this have calculated the field in the corona.

Besides radiating light and radio waves, the sun also emits some X-rays—the same type of radiation used in medical diagnosis. SkyLab took X-ray photographs of the sun that revealed the appearance of "holes" in the corona. The magnetic field data were analyzed for the days on which the photographs were taken, and when the results were compared with the photographs, they showed that the magnetic lines of force seemed to diverge from these coronal holes (Figure 4.5).

The complex relationship between solar activity and the magnetic fields of the sun, and the various theories that have attempted to explain this relationship, as well as the methods used by different research groups in their attempts to predict when solar maxima will occur, is an extensive area of study, so we will leave it to later in this chapter.

The Solar Wind and the Interplanetary Magnetic Field

The solar wind is the continual, almost radial, outflowing of the solar corona. The coronal temperature is so high that the gravitational field of the sun cannot contain this highly ionized gas in a confined static atmosphere.

Two components of the solar wind have been identified: one, a slow-moving low density flux, and the other, superimposed on this, are high-speed streams. This latter emission is greatly increased over coronal holes.

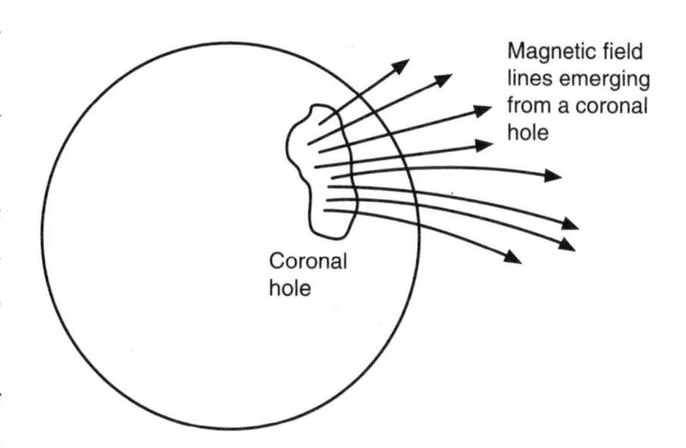

Figure 4.5
Magnetic field lines diverging from a coronal hole.

The loops of the solar magnetic field that arch into the corona and are anchored in the active regions of low latitudes will be stretched out, near the ecliptic, into interplanetary space by the outflowing solar wind. While these loops are pulled out radially, the sun rotates about its axis, leading to the winding-up of the stretched-out field lines. As a result the interplanetary field assumes a spiral configuration. This is called the *garden hose effect*, because the lines of the magnetic field, similar to the water jet of a rotating garden hose, form a curved spiral, but the solar wind particles, similar to the droplets of water, always move out in a radial direction (see Figure 4.6). Although the particles of the solar wind move away radially, an observer on Earth has the impression that they are coming from a direction approximately 5 degrees to the west of the sun. This is due to an aberration effect produced by the motion of Earth at right angles to the sun-Earth direction. The effect is the same as the case of a moving observer who thinks that the rain is falling at an angle, while a stationary observer sees the rain falling vertically (see Figure 4.7). Besides having a spiral form, the interplanetary field is divided into four

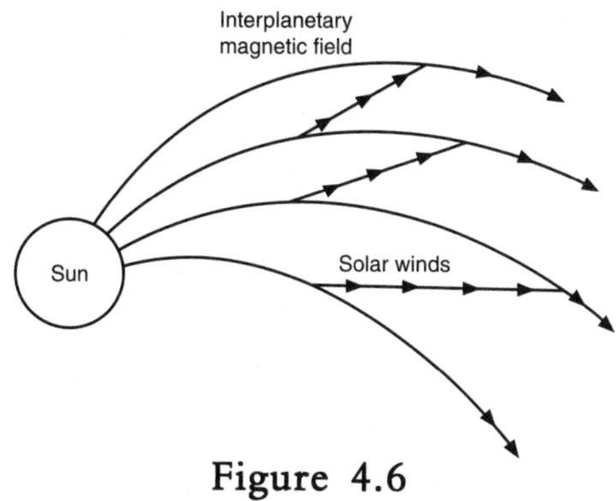

Figure 4.6
The solar wind and the interplanetary magnetic field.

Figure 4.7.
The aberration effect.

sectors, with the field pointing in different directions in alternate sectors (see Figure 4.8).

Associated with the interplanetary magnetic field is a sheet of electric current. The warping of this sheet is due to the sectorial division of the solar magnetic field in the equatorial plane of the sun. A three-dimensional model of the sheet is shown in Figure 4.9. The component of the interplanetary field that is perpendicular to the ecliptic, changes sign when the Earth passes through the current sheet as it orbits the sun.This can lead to a reconnection between the lines of the interplanetary field and geomagnetic field in the northern hemisphere, though not in the southern hemisphere, where the interplanetary field has a downward component. When this happens, charged particles can more easily enter the Earth's ionosphere, because charged particles preferentially travel along field lines. Geomagnetic activity thus changes significantly depending on which sector the Earth is in.

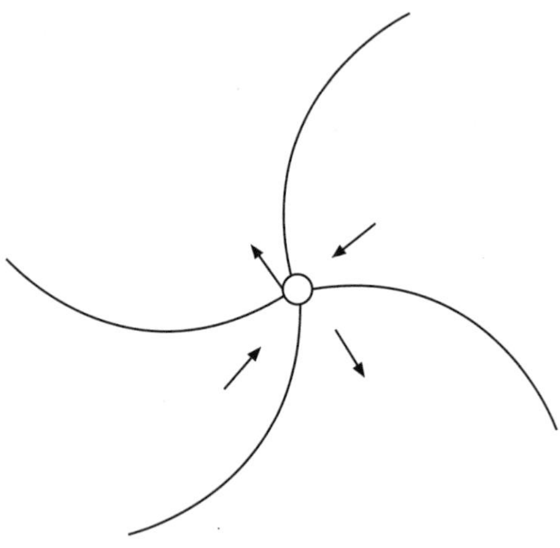

Figure 4.8
Sector boundaries of the interplanetary field.

Activity on the sun can also modulate the charged cosmic ray particles that come from beyond the solar system. One well-known way in which this can occur is called the *Forbush decrease*. Professors Thomas Gold and Eugene Parker both proposed theories to account for this sudden decrease in extrasolar system cosmic ray flux, which follows a large flare event on the sun. According to Gold, a plasma cloud ejected by a large flare event can pull out the solar magnetic field to form a large magnetic "bottle." Such a bottle can eventually engulf the Earth and thus provide an additional shield from cosmic ray particles (see Figure 4.10). Parker suggests that a large flare event gives rise to a blast wave (similar to a sonic boom), in which the strength of the interplanetary field is increased. The blast wave propagates out into space, and when it reaches the Earth it provides additional protection against cosmic rays. The abrupt increase in the interplanetary magnetic field within a blast wave has been observed by some spacecraft, thus providing support for Parker's theory (see Figure 4.11).

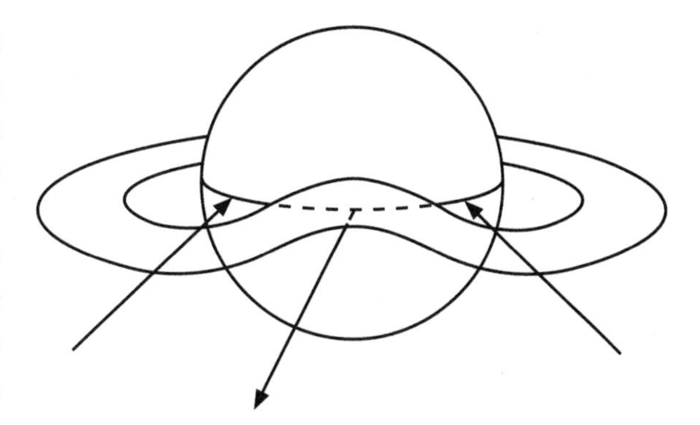

Figure 4.9
Three-dimensional model of the current sheet.

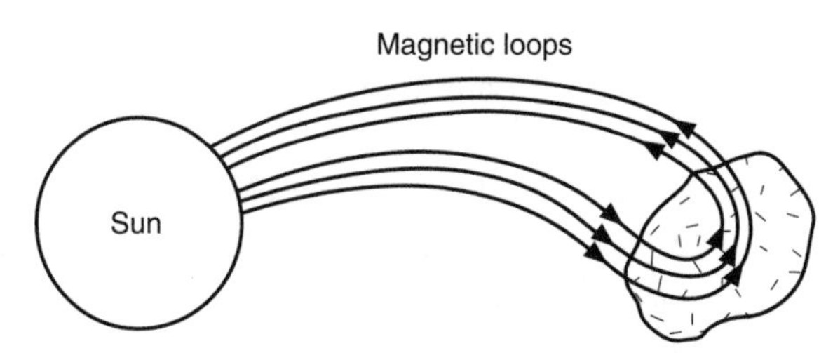

Figure 4.10
Gold's model to explain the Forbush decrease.

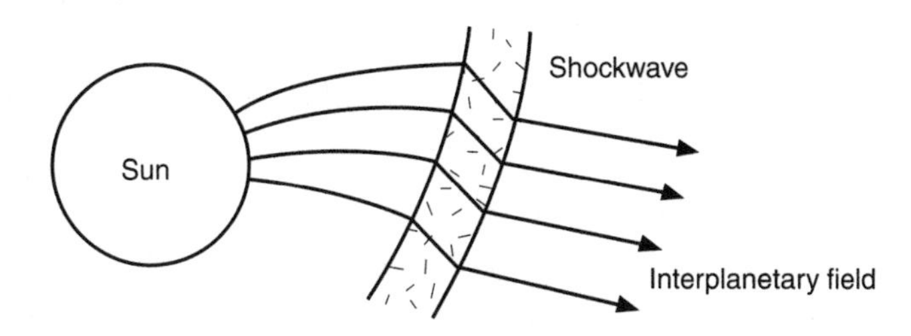

Figure 4.11
Parker's model for the Forbush decrease.

SECTION TWO—UNDERSTANDING THE SOLAR CYCLE

In this section we examine the various attempts to explain the solar cycle. Although some technical details are introduced, we examine the various theories from the point of view of the philosophy of science. We also claim that this is one of a few cases, in modern physics and astronomy, in which there is a conflict between the philosophy of applied mathematics and the philosophy of science. Whereas applied mathematicians are more committed to their mathematical protocols, which have proved useful in other branches of theoretical science, astrophysicists are more committed to other types of models that are able to explain a wide range of observational details relevant to the solar cycle.

The Maunder Butterfly Diagram and the Hale-Nicolson Laws

Richard Christopher Carrington (1826–1875) was an English amateur astronomer who discovered, in 1863, that the material near the equator of the sun was rotating faster than the material near the poles. Carrington concerned himself with the sunspot distribution over the surface of the sun, and concluded that new spots appear at higher latitudes, and migrate toward the equator, where they eventually disappear. The systematic progression of sunspots toward the equator is called *Sporer's law*, because it was Friederich Wilhelm Gustav Sporer (1822–1895), a German astronomer, who, in 1861, combined all available records to produce a statistically significant result. Edward Walter Maunder (1851–1928), an English astronomer, was responsible for introducing the Butterfly diagram, which is a visual representation of the movement of sunspots and Sporer's law.

George Ellery Hale, in 1908, reasoned that the splitting of sunspot spectral lines was a manifestation of the then-recently discovered Zeeman effect, which showed that magnetic fields will lead to a splitting of the spectral lines of atomic elements. This led Hale and his coworkers to the discovery of the Hale-Nicolson laws:

1. All pairs of sunspots in the same hemisphere have opposite polarity with regard to the leading and trailing members of the pair (if the leading member is a south magnetic pole, then the trailing member will be a north magnetic pole), but the polarities will be opposite from those in the other hemisphere.

2. During the next solar cycle the polarities in the two hemispheres will be reversed.

This second law implies that the solar cycle is roughly 22 years long, rather than 11.

Babcock's Model of the Solar Cycle

One of the first models that could explain, in qualitative terms, the most important observations of the solar cycle was that proposed by Babcock. Here is a summary of its main features, which can best be described in terms of a few stages.

Stage One: During sunspot minimum, the field resembles the dipole field of a simple bar magnet, with the field lines entering near one pole of the sun and emerging from the other pole.

Stage Two: Differential rotation of the sun means that its rotation rate increases as one moves from the poles toward the equator. This type of rotation means that the field lines near the equator will be stretched into toroidal loops—magnetic lines of force that circle the sun, nearly parallel to the solar equator, resembling the parallels of latitude on Earth—and small-scale convective motions twist these loops into a braided configuration, resembling ropes: both processes serve to intensify the field strength.

Stage Three: Because of the magnetic pressure in these ropes, the plasma density will be lower in any one of these ropes than it is just outside the rope. This will make sections of the flux rope buoyant, causing these sections to raise to the surface to form a loop prominence, the feet of which will form a sunspot pair.

Stage Four: This describes the subsequent reversal of the field. Observations show that the trailing members of sunspot pairs migrate toward the nearest pole, while the leading members migrate toward the equator. The lines of force emerging from the trailing members will initially neutralize the older field near the poles, and subsequently replace them with lines of opposite polarity.

Stage Five: Differential rotation will give rise to a straightening-up process, which will give a new dipole field with opposite polarity.

The Babcock model gives a reasonable qualitative description of the observations.

Mathematical Models of the Cycle

Several researchers have attempted to construct mathematical models of the solar cycle, using the general equations of magnetohydrodynamics. The major obstacle to this approach is the nonlinearity of the equations. This means that if you change the flow of the plasma you also change the structure of the magnetic field, and vice-versa. There is no simple mathematical way of solving these equations in a way that is rigorous and consistent. This is why all workers in the field have used what is known as the *kinematic approach.* In this approach, one starts by ignoring the effect of the magnetic field lines on the plasma, and one then works out how the motions of the plasma will distort the field lines. During stage two, one calculates how the distorted field will affect the movement of the plasma. Then one works out how the modified flow will further distort the field lines. In essence, it is an iterative process. Although some of the applied mathematicians working on the problem have convinced themselves that progress is being made, many researchers in this branch of research remain skeptical of the whole approach.

The mathematical problems with this particular approach were highlighted by Douglas Gough, an applied mathematician working at Cambridge University in England, in a letter to *Nature* in 1990: "Subsequently it became fashionable to consider the cycle as the product of a turbulent dynamo confined solely to the outer regions of the Sun, where convection occurs. The idea fell out of favor when detailed calculations failed to explain the most salient features of the observed sunspot distribution." Gough then says that attention is being turned to the interface between the convective zone and the radiative interior, in the search for a theory of the solar cycle. He speculated that, "both magnetic activity and structural changes in the superficial layers of the sun are merely symptoms of a more deeply seated dynamical process," but admits that, at the moment, it is still not clear what is the underlying mechanism that controls the solar cycle.

In 1993, K. Moffatt drew attention to the problems with relying on buoyancy theory to carry parts of the flux tubes to the surface: "...the simplest theory of magnetic buoyancy...suggests a much shorter timescale [shorter than the 22-year solar cycle period] of the order of months or less...so short that a dynamo process located in the convective zone appears to be at variance with observed solar activity."

The main problem with mathematical models is that they are riddled with assumptions that are impossible to justify. In order to investigate mathematical models, either using algebraic formulae or computers, assumptions have to be made about the scale of the irregularities in the magnetic field, how these irregularities affect the movement of the plasma, the effect of turbulent motions in the convective zone on the field itself, and the stage at which it becomes necessary to consider the effects of amplification of the field on the movements of the plasma. The first makes use of a differential rotation model that best fits the observations, and then works out how this will distort the field lines. Those applied mathematicians who promote this type of approach then work out how the strength of the subsurface magnetic field changes with time, solar latitude, and longitude. In the second stage, the emergence of the field due to buoyancy is then considered as a separate problem, and the assumption is made that more of the field will emerge where the subsurface field is at its strongest. Many of these models are presented at conferences before they are published, and the titles of the conference presentations and the published collected papers carry misleading phrases such as, typically, "recent progress in understanding sunspots and solar flares," or "recent advances in understanding the solar cycle." In all cases the progress is in handling the mathematical models, and the results do not lead to a better understanding of the astrophysics of the situation or to practical methods of predicting when sunspots and prominences are likely to occur. All these models also ignore the evidence that there are links between planetary alignments and particularly violent events in the solar cycle.

I will end this section with two comments which highlight the mathematical difficulties involved. The first is from an article by C. Jones, called "Making Sense of a Turbulent Universe," which appeared in *Physics World* in 1995: "Despite these [mathematical and computational] difficulties...the dynamo problem has stimulated mathematicians, physicists, astronomers, and geophysicists. As with other 'grand challenge' projects, it is the ideas developed along the way that may in the end prove more successful."

The second comment was made 10 years later. It is from an abstract to an article by Wilmot-Smith, Martens, Nandy, Priest, and Tobias, called "Low-Order Stellar Dynamo Models," which appeared in *Monthly Notices of the Royal Astronomical Society* in 2005: "Stellar magnetic activity—which has been observed in a diverse set of stars including the sun—originates via a magnetohydrodynamic mechanism working in stellar interiors. The full set of magnetohydrodynamic equations governing stellar dynamos is highly complex, and so direct numerical simulation is currently out of reach computationally."

Other Models That Attempt to Predict Aspects of Solar Activity

I want to point readers toward a review of predicting the solar cycle in a paper by Dr. David Hathaway, of the Science Directorate Marshall Space Flight Center, called "Solar Cycle Prediction," which was posted on the Web on April 2, 2008. I leave those interested in the details to look at this Web page (see the Bibliography), but I want to make a few comments of my own about the methods used by the researchers quoted by Hathaway, and to express my own objections to such approaches.

Generally, these scientists do not attempt to model the solar magnetic field and its evolution, either mathematically or in detailed physical terms. Instead they rely on making a detailed statistical analysis of past cycles, relating these to current solar activity, and then projecting their work into the future to give some indication of the time and amplitude of the next solar maximum. As we have already pointed out earlier in this chapter, the fact that solar activity is connected to changes in the Earth's own field—the geomagnetic field—had already been established by the end of the 19th century. This fact has been used by the researchers mentioned by Hathaway, so they not only restrict themselves to the study of past solar activity, but they also make use of the linking between solar activity and geomagnetic activity. This means that their methods have shed little light on the solar subsurface mechanism that gives rise to sunspot activity, they have not clarified how or why the Maunder Butterfly diagram arises, and they have ignored the work of those scientists who worked on mathematical models of solar activity.

Furthermore, the mathematical computer modelers, and the empirical data computer modelers, all have totally ignored the fact that there is considerable evidence to show that the movements and positions of the planets do have an effect on the solar cycle and on violent events on the sun.

Evidence for Planetary Influences on the Solar Cycle

In my book *Cosmic Magnetism*, I reviewed some of this evidence, and speculated on how one might explain the observed correlations. There I said: "There are two possible ways in which the planets could affect the solar activity. First, the sun's movement around the common center of mass of the solar system could cause changes in the pattern of the convective motions that generate the large-scale magnetic field, and this in turn could lead to large-scale reversals of the field. Tidal action could then foster the growth of instabilities in those regions where the magnetic field configuration is in unstable equilibrium."

In 1986 I suggested to Andrew Turner, who was then one of my third-year students, that he might like to look at my suggestion in more detail, for his final year project. He was only too pleased to do so, and in June 1987 he presented his excellent thesis on the subject. Not long after this, a friend of mine, Michael Willmott, who had studied chemistry and computer science at university, and at that time worked as a computer specialist at a commercial firm, asked if he could do a PhD at the University of Plymouth. He was also a very keen amateur astronomer and wanted to do a project that combined his computer experience with astronomy. I asked him if he wanted to continue with the work started by Turner. He was very excited by the idea, and so he soon set to work on an extension of Turner's project. When the British Association for the Advancement of Science met in Plymouth in August 1991, Willmott had made sufficient progress for us to present a joint paper at this meeting.

Peter Beer, one of the editors of *Vistas in Astronomy* (an international review journal for astronomy and astrophysics), and also son of Arthur Beer, who was the founder of the journal, was present at this meeting. After we had presented our paper, he asked us to write an extended paper on it for his journal. The paper, entitled "Sunspots, Planetary Alignments and Solar Magnetism" (SPASM for short) appeared in 1992. Mike Willmott continued his calculations, and in the Winter 2000/2001 *Journal of the National Council for Geocosmic Research* we published a joint article called "A New Theory for the Solar Cycle." Much of what follows is based on the joint research I did with Willmott, although Turner did some of the original calculations.

Between 1859 and 1975, a great deal of research was carried out on links between solar activity on the one hand, and planetary movements and alignments on the other hand. For those interested in this extensive literature, I refer the reader to my paper with Willmott and Turner in *Vistas in Astronomy*, where we have undertaken a detailed review of the literature and give full references. In this section we just wish to highlight some of the findings that are particularly important to the theory discussed later in this chapter.

In 1965, P.D. Jose showed that there was a correlation between the sign of the rate of change of angular momentum about the common center of mass of the solar system and the sunspot cycle (see Figures 4.12(a) and (b)). He had in fact calculated the rate of change of angular momentum of the sun about the instantaneous center of curvature of its orbit about the common center of mass of the solar system. The turning points in this curve correlated well with the sunspot number curve.

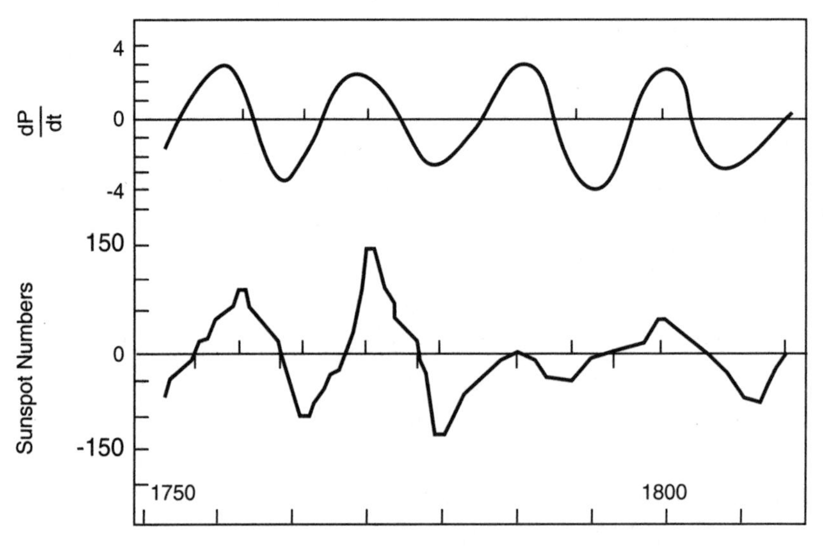

Figure 4.12 (a)

Movement of the sun about the common center of mass of the solar system and the sunspot number curves (1750 to 1800-plus).

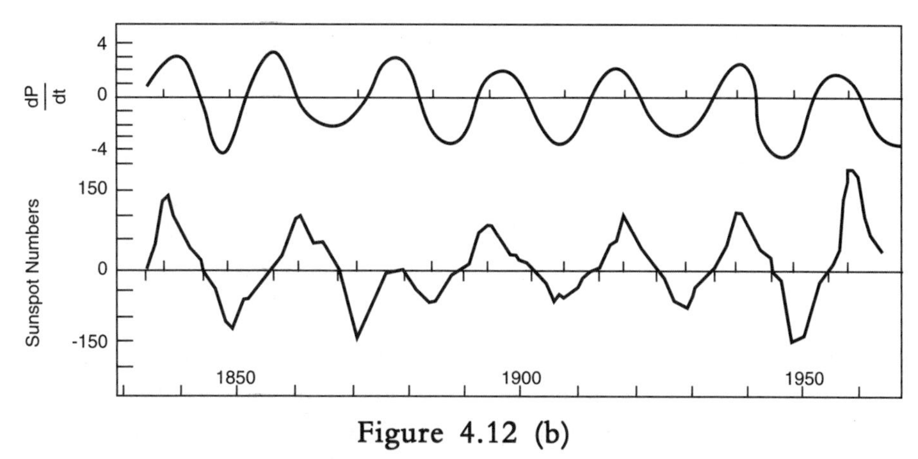

Figure 4.12 (b)
Movement of the sun about the common center of the mass of the solar system and the sunspot number curves (1800 to present day).

I now turn to the important points made by Blizard in her "Long-Range Solar Flare Prediction" in 1969:

- Planetary positions, notably conjunctions, are associated with increased solar activity, as typified by the following: the areas of large sunspots correlate strongly, and x-rays and radio emission correlate to a lesser extent, with significant planetary conjunctions; the activity during the period March to April 1966 appears to be related to an alignment of Saturn, Uranus, and Pluto; successful correlations between conjunctions and "jerk" (rapid changes in the movement of the sun about the common center of mass of the solar system) with solar activity features, such as proton events and ground-level events, were achieved for cycles 14 to 20; conjunctions of Mercury, Venus, and Jupiter seem to be the trigger mechanism for proton events through tidal effects and jerk; only three combinations of the positions of the planets Mercury, Venus, Earth, and Jupiter correlated with two-thirds of the proton events recorded between 1956 and 1961, corresponding to a probability of 0.0005 by Chi-squared analysis (the correlation was highly significant in statistical terms!).

- When considering planetary positions in relation to solar activity, factors such as active (solar) longitudes, the phase of the 11-year cycle, and relative activity of the cycle need to be contemplated.
- On studying solar activity in relation to (solar) latitude, it was found that the angular distance of planets above or below the solar equator affected the north-south distribution of spots and flares.
- Initial tidal effects are extremely small, but the result in the corona is an amplified waveform due to the steep gradients of temperature and density prevailing. (According to Blizard this mechanism could possibly play a significant role in influencing solar activity.)
- The reason for the inconsistency of correlation between dynamic parameters such as the rate of change of angular momentum, tidal force squared, and the rate of change of tidal force on the one hand, and solar activity on the other hand, is that two of these parameters may be involved in some nonlinear way.

All of Blizard's observations are readily explained by the Seymour-Willmott theory of the solar cycle, which will be described in detail later in this chapter.

A great deal of the work done on possible planetary influences on the solar cycle used what is called the *equilibrium theory of the tides.* This theory tells us that the difference between high tide and low tide should be no more than a few feet, which works reasonably well for the open oceans, but it fails completely for bays and estuaries, ports and harbors, where the sea interacts with the land; in these situations, the tides are generally much larger. Some places have particularly large tides, which serve as examples of tidal resonance. Let us explore a few of these examples.

In the Bay of Fundy, in Canada, the tidal range is 46 feet; in St. Malo, in France, it is 33 feet; and in the extreme north of the Gulf of California, in Mexico, it is 26 feet. How does this happen? To explain this we need to introduce a new scientific term—a *seiche.* To start with, every body of water has a number of natural frequencies, or resonances. For example, if you rock back and forth in a bathtub you can easily get the waves to grow until they overflow the bath—this resonant oscillation of the water is a seiche. If one sets off a small explosion at the mouth of a bay, it will create a wave that will travel up the bay and back. The time it takes this wave to travel up and down the bay is called its *natural period* or *seiche period*, which depends on the length and shape of the bay. It is not dissimilar from blowing air across the end of an organ pipe: This will cause a vibration in the pipe, which is the natural period of the pipe; the longer the pipe, the longer will be its natural period. Earthquakes can cause seiches in swimming pools (during the Northbridge earthquake of 1994, swimming pools all over Southern California overflowed, and during the great Alaska earthquake of 1964, swimming pools as far away as Puerto Rico were set into oscillation). Tides in marginal seas and bays cannot be directly forced by the tidal effect of the moon; they are co-oscillation tides generated by the tidal movement at the connection with the ocean basins. If the tidal force is in resonance with a seiche period of a bay, the tidal range is amplified, and it can be enormous.

How can all this be relevant to the sun, where there are no land masses? It is known that magnetic fields can constrain very hot ionized gases, called *plasmas*. Indeed, current attempts to generate power using the type of thermonuclear reactions that occur deep in the interior of the sun use strong magnetic fields to squeeze such hot gases. Strong magnetic fields that develop beneath the surface of the sun, late in the solar cycle, take the form of canals that are parallel to the solar equator, and these can constrain the flow of plasmas in the sun. We will discuss this topic in more detail later in this chapter.

The Use of Analogous Models in Theoretical Physics

The history of theoretical physics provides one with many interesting insights into the trials and tribulations that have to be faced by those who seek to give mathematically quantifiable descriptions of natural phenomena. Our increased understanding of atomic physics provided a relevant specific example: The physicists J. Fraunhofer, G.R. Kirchhoff, and R.W. Bunsen laid the observational and experimental foundations of atomic spectroscopy, which enabled scientists to identify chemical elements by the spectral lines they emitted or absorbed. It was soon after this that G. Stokes, W. Thompson, B. Stewart, and G. Kirchhoff suggested that certain resonant frequencies characterized each element. This idea was further developed by the Dutch theoretical physicist H.A. Lorentz. Although at this time scientists had virtually no concept of what the atom looked like, it was known that electric current was made up of charged particles. Lorentz applied the theory of damped simple harmonic motion, the same theory used to discuss the swinging of a pendulum in a liquid, to the oscillations of electrons within the atom. He was able to explain some of the basic properties of the interaction between electromagnetic radiation and the atoms of specific elements. In this vein I would like to quote from *Introduction to Theoretical Physics* by J. Slater and N. Frank: "In optics, the theory of refractive index and absorption coefficient is closely connected with resonance. As is shown by the sharp spectrum lines, atoms contain oscillators capable of damped simple harmonic motion, or at any rate they act as if they did: the real theory, using wave mechanics, is complicated, but leads essentially to this result."

We thus see that Lorentz's simple theory provided a basis for identifying atoms from their spectra, thus enabling their physical and chemical properties to be deduced. This was more than a decade before Bohr, Sommerfeldt, Schrödinger, Heisenberg, and Dirac developed quantum mechanics, which allowed us to understand these characteristics in terms of the electronic structure of atoms.

The analogous model I am proposing for the solar cycle is able to explain several aspects of solar activity, including the correlation between planetary movements and alignments and violent events on the sun.

A New Model for the Solar Cycle

I will start by assuming that the normal ideas of dynamo theory can be applied to the initial stages of the cycle. According to dynamo theory, a basically dipolar field is distorted, by differential rotation, into a field with a substantial toroidal component that becomes more prominent near the equator. Small-scale cyclonic motions twist loops into this field, and the loops combine to produce the new dipole field. Willmott and I suggest that near the poles, sometime between solar maximum and minimum, the small-scale cyclonic motions are reversed, thus giving rise to a field of opposite polarity. The reversal is caused by the movement of the sun about the common center of mass of the whole solar system. The principal evidence to support this proposal is the work of Jose, who found evidence linking the sunspot number curve with the movement of the sun about the instantaneous center of curvature of the sun's orbit about the center of mass of the solar system.

Near the poles, the sun spins on its own axis once in about 33 Earth days. Jupiter is moving about the common center of mass of the solar system roughly about once every 12 years. At maximum displacement, the center of the sun is about 2 solar radii from the center of mass of the solar system, and at such times the material within 2 degrees (of latitude) of the poles has two comparable coriolis effects acting on it—one due to solar rotation, and one due to the movement of the sun about the common center of mass (of the solar system), due largely to the seesaw effect of Jupiter. It is known from observations that the dipole field starts to change polarity in the polar cap region, so we would expect this to occur when the pole of the sun has a maximum displacement from the center of mass. This is in keeping with Jose's observations.

When the sun's magnetic field begins to emerge in the form of sunspot pairs linked by solar prominences, the field lines in the latitudes at which these phenomena occur are very nearly parallel to the equator. Willmott and I made the assumption that, at this stage, it is more appropriate to use Airy's canal theory of the tides (explained in detail in the next paragraph) to discuss possible planetary effects on the formation of sunspots, because we are interested in the interaction of planetary tidal forces of the planets and the canal-like structures beneath its surface, rather than their effect on the whole sun. The emergence of the magnetic tubes of flux are normally attributed to magnetic bouyancy, but, as already noted, buoyancy is embarrassingly effective, leading to a solar cycle that is far too short. This difficulty can be avoided if we assume that the tubes of flux have been twisted in largely force-free, helical-like magnetic structures by the small-scale cyclonic motions in the convective zone of the sun. The properties of such force-free fields were investigated by G.F. Freire in 1966. The twisting of the field lines would considerably reduce the effects of buoyancy, and the helical-like magnetic tubes of force would be in a state of unstable equilibrium. The tidal forces of the planets, amplified by tidal resonance, could easily provide the necessary additional impulse to cause such structures to rise to the surface.

In 1845, George Biddell Airy, who was the Astronomer Royal at the Royal Observatory in Greenwich at the time, formulated a theory that could, mathematically, deal with waves in an imaginary canal encircling the Earth, parallel to the equator. Earlier in this chapter we discussed how tides in bays and estuaries could be very large, if the natural period (or seiche period) of the bay was close to the tidal period of the open ocean, which was caused by the gravitational tug of the moon on the oceanic waters. Airy introduced the concept of the "free wave" for such a canal, which is the speed with which a wave would naturally propagate along such a canal. This speed depends on the width and depth of the canal. Airy showed that if the dimensions of the canal were so chosen that the speed of the free wave traveled with the same speed with which the sublunar point (the point on the Earth's surface immediately below the moon) traveled over the Earth, then the tides in the canal would be very large indeed. If the speed of the free wave was equal to, or greater than, the speed of the sublunar point, then the tide would be direct—that is, the point immediately below the moon would have a peak. If the speed of the free wave was less than that of the sub-lunar point, then the tide would be inverted—that is, there would be a trough immediately below the moon.

The seiche period of a circular canal would be the circumference of the canal divided by the speed of the free wave. The circumference of such a canal would be latitude dependant—that is, the farther the canal was from the equator, the smaller would be the circumference, and hence the seiche period would be shorter. Willmott and I applied these ideas to the canal-like structures that developed in the convective zone of the sun as the solar cycle built to a maximum.

A stretched magnetic line of force is rather like the string of a violin, in that if it is plucked, it will vibrate. If one had a long bundle of such strings, then plucked one end, the disturbance would propagate to the other end with a speed called the *Alfven speed*. This speed depends on the strength of the magnetic field in the bundle, and the density of matter in which it was embedded. It is known that the magnetic fields that emerge from the poles are about 1,000 times weaker than those measured when the sunspots emerge. The amplification of the field is due to differential rotation winding up the lines of force (see Figures 4.13, 4.14, 4.15, 4.16, and 4.17), and also due to the convective cells twisting the lines into the force-free ropes we mentioned

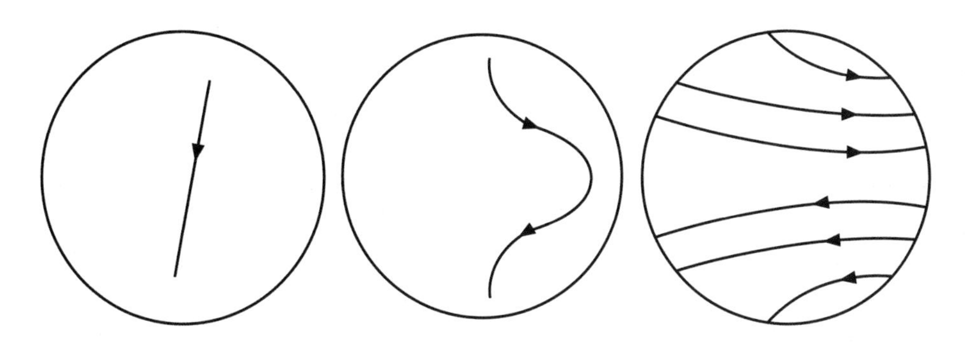

Figures 4.13, 4.14, 4.15
The winding-up of the solar magnetic field due to differential rotation of the sun.

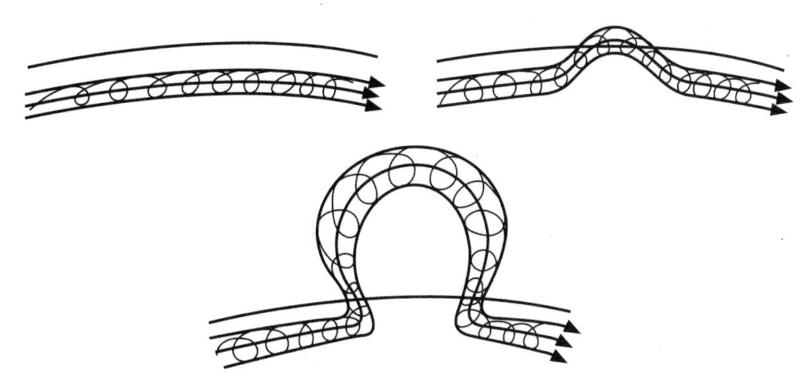

Figures 4.16 and 4.17
The twisting-up of the field lines to form helical fields.

earlier. Thus, as the solar cycle progresses, the strength of the field will increase, and the Alfven speed in the canals will increase. This means that the seiche periods for canals in middle latitudes will be shorter than those nearer to the equator, and thus they will resonate with the tidal frequencies of the planets before those closer to the equator. As the Alfven speed increases, it then becomes possible for canals closer to the equator to resonate with the tidal periods of the planets. Our theory thus provides a natural explanation for why sunspots migrate toward the solar equator as the solar cycle progresses.

Bear in mind that it is the speed of a planet with respect to an element of matter in these canals that is important, and from the point of view of such an element, the speed of Mercury will be slower than that of Saturn. This can be further clarified by looking at an element of matter in a canal, at an intermediate solar latitude, where the sun is rotating once every 30 Earth days. Mercury is going around the sun once every 88 Earth days, so that in 30 days, Mercury has moved through about one-third of its orbit, and the element of matter in the canal has to go a bit further to catch up with Mercury, so from its point of view, Mercury is taking closer to 40 days to go once around the sun. However, in 30 days, Saturn, which has an orbital period of 29 Earth years, will not have moved very much, and the element of matter will think it is going around the sun in about once every 30 days. This means that, early in the solar cycle, the inner planets will be more important to resonant tidal disruption than the outer planets.

Blizard, in her researches, had found a result she could not explain: sometimes, when two planets were at 90 degrees with respect to each other, there were violent events on the sun. This is contrary to what one would expect if one generalized from what happened with oceanic tides: At new moon and at full moon, when the sun, Earth, and moon were in a straight line, then we got the spring tides, which were the highest tides of the lunar month. At first quarter and last quarter of the moon we have the neap tides, the lowest tides of the lunar month, and at such times the sun and moon make angles of 90 degrees with each other, as seen from Earth. This is only true for the equilibrium theory of the tides, which applies to the open oceans and the shelf seas. Airy's canal theory of the tides showed that if one had to choose the dimensions of a canal parallel to the equator, such that the speed of the free wave was greater than that of the sublunar point but less than that the subsolar point, then in such a canal one could get spring tides when the sun and moon made 90-degree configurations with respect to each other as seen from Earth. The same reasoning can be applied to the magnetic canals on the sun, which exist as the sunspot cycle builds to a maximum. If the Alfven speed in such a canal is greater than the subplanetary point for one planet, but less than the subplanetary point for another planet, then a 90-degree configuration of these two planets will give rise to violent events on the sun, which was an aspect of solar activity planetary alignment links already noticed by Blizard.

Willmott, Turner, and I showed that for resonant tidal reactions between the subsurface solar magnetic canals and the gravitational tug of the planets to occur, the Alfven speed has to be very nearly equal to the speed of the subplanetary point of one planet or another. Because it is simple to calculate the subplanetary point speeds of the planets with respect to elements of matter in canals at various solar latitudes, we can calculate the required Alfven speed for resonance to occur. As we have already noted, the Alfven speed depends on the strength of the magnetic field and the density of matter within a magnetic canal. However, because we know the strengths of the magnetic fields within sunspots (measured using the Zeeman effect), we can calculate the density of the material needed for resonance to occur. Our calculations showed that the required density would be found in the convective zone of the sun.

The Advantages of the New Model

- The new theory is able to explain why certain planetary alignments, as seen from the sun, give rise, at various stage in the solar cycle, to violent events on the sun.
- It also explains the evolving distribution of sunspots, as the cycle progresses, which is embodied in the Maunder Butterfly diagram.
- The theory is consistent with solar activity taking place in the convective zone of the sun.

Part II

The Electrified Tunnels of the Cosmic Tapestry

Introduction to Part II

Formulating a New Theory of Matter, Space, and Time

The 20th century saw a host of major achievements in our understanding of the physical universe. This progress made it clear that 19th-century physics, now known as *classical physics*, had severe limitations. The rise of modern physics did much to come to grips with these problems, but it was to encounter severe problems of its own. I want to give an overview of the difficult transition between classical and modern physics. (More details will be given in Chapter 6.)

Astronomy Before 1900

Astronomy before 1900 was based mainly on what we could see. Some astronomers did believe that there might be other forms of radiation linking us to the universe, but there was very little evidence available to support this belief. Classical physics, based largely on gravitation and electromagnetic theory, led physicists and astronomers into conflict with biologists and geologists, with regard to the age of the Earth. The naturalist

Charles Darwin (1809–1882) had formulated the theory of evolution by natural selection, which required a longer time-span for the age of the Earth than that estimated on the basis of classical physics. Geologists also wanted a longer time-scale for the evolution of our Earth. At the end of the 19th century there was a stalemate between the physical scientists and the earth and biological scientists. The resolution of this situation had to await the rise of modern physics in the first two decades of the 20th century.

The Rise of Modern Physics

In 1900, Max Planck (1858–1947) initiated the birth of modern physics by formulating the quantum theory of radiation, in which the wavelength at which a hot body emitted most of its radiation was known to be related to its temperature. Thus relatively cool bodies gave off most of their radiation at the wavelengths associated with the color red, very hot bodies gave off most of their radiation at blue wavelengths, and bodies at intermediate temperatures, such as our sun, gave off most of its radiation at yellow wavelengths. Planck showed that energy from hot bodies can only be emitted in indivisible amounts, called *quanta*, the magnitudes of which are proportional to the frequency of the radiation.

James Clerk Maxwell, whom we already mentioned in Chapter 2, had shown that in free and open space, all types of electromagnetic radiation traveled at the speed of light. The speed of light became a cornerstone of Albert Einstein's theory of relativity. In 1905 he formulated the special theory of relativity, which showed quite clearly that nothing could travel faster than light. Einstein also showed that Planck's concept of quanta could be used to explain how certain wavelengths of light could be used to eject electrons from particular types of metals, under specific conditions.

Ernest Rutherford (1871–1937), with his work on radioactivity, and his subsequent work on the atomic structure, was able to resolve the conflict between the physicists and astronomers, and geologists and biologists, by showing that there was another form of energy, now known as *nuclear energy*, which was not known to classical physicists. When applied to our Earth and the sun, it gave a much longer time-scale for the evolution of these bodies.

Magnetic Fields in Astronomy

The 20th century also saw great developments in our understanding of the widespread nature and importance of magnetic fields in astronomy. Optical astronomers, their radio-astronomical colleagues, and the discoveries made by space scientists using satellites and space probes, showed us that our sun, some of the planets, many different types of stars, our Milky Way galaxy, and several other galaxies, had magnetic fields. There is also evidence to suggest that there are magnetic fields in clusters of galaxies, and there might well be magnetic fields between galaxies and clusters of galaxies. Thus magnetism was another link between us and the rest of the universe. These fields could influence the particle and magnetic environment of our Earth. The details of these fields, their interactions and activities, were discussed in some detail in the first part of this book.

Further Developments in Modern Physics

As our understanding of the atom and quantum theory began to develop further in the first 30 years of the 20th century, the first signs of a conflict between relativity and quantum theory also began to appear. Bohr applied quantum principles to the structure of the atom proposed by Rutherford, and this led to a model for the atom in which the tiny electrons circled the much heavier atomic nucleus in certain allowed, stable orbits. When an electron was in one of these stable orbits, it would radiate electromagnetic waves; however, if an electron absorbed light, or another form of radiation, it would move from an orbit closer to the nucleus to one further out. When it fell back to its origin orbit closer to the nucleus, it would radiate the radiation it had absorbed.

The next step forward was made by Prince Louis de Broglie (1892–1987), a French nobleman. Einstein had shown that waves can sometimes behave as if they were particles. This idea was inverted by de Broglie, who claimed that particles could show a wavelike nature. This led to the realization that the allowed Bohr orbits were exactly those into which one could fit a whole number of de Broglie waves associated with electrons moving with a given speed. The waves that guided the electrons were not electromagnetic waves; they were called *matter waves*, and for this idea to work they had to travel faster than light, thus violating the basic idea of Einstein's special theory of relativity. (More details on these ideas will be given in Chapter 6.)

Bell's Theorem

John Stuart Bell (1928–1960) formulated a theorem, now known as Bell's Theorem, that was to highlight how strange quantum physics really was. In essence, this theorem shows that if two subatomic particles were created at the same point and at the same time in a laboratory and then subsequently separated, then measurements on the two particles will always be correlated. Such subatomic particles spin on their own axes, somewhat similar to the way the Earth spins on its axis. If the one particle was spin up (spinning on its own axis in a clockwise direction as seen from above), then the other particle will be spin down (spinning in an anti-clockwise direction as seen from above), and this situation will remain so, no matter how far apart they are. What is more, if one had to change the spin of the one particle, a message of the change will get to the other *faster than the speed of light.* Einstein referred to this situation as "spooky action at a distance." Modern physicists call it *entanglement,* because the histories and subsequent behavior of the particles are entangled in a very intimate way. The effect does not weaken with distance. Two possibilities exist: Either quantum information can travel from the one particle to the other, almost instantaneously, or the two particles should not be seen as separate entities.

The Plasma Space Theory of Matter

In Chapter 6, I proposed a new theory of matter. In this theory all particles have mass because they have electric lines of force wound up in them, and their masses arise from the electrical energy stored up in them according to Einstein's formula $E = mc^2$. The neutron is the basic starting particle, but it is unstable when it is outside the nucleus, and decays into a proton, an electron, and an anti-neutrino. The cores of the electron, the proton, and the neutrino are force-free windings of the electric lines of force, which means that they have no tendency to kick apart, or unwind. However, the neutron has an outer cloak of lines of force, which is unstable when it is out of the nucleus, so when it decays, this outer clock rearranges itself, so that in ordinary space-time, the lines of force go from the proton to the electron.

But in plasma space, the lines of force go from the electron back to the electron sheathed in tubes of insulating space, so we are not normally aware of them, just as we are not aware of large electric currents passing through insulated electric wires. All particles, whether they have charges or not, can be pushed around by the incessant motion and vibrations of these tubes of insulating space, and it is this that gives rise to the de Broglie "pilot waves" or to the fluctuations of David Bohm's "psi-field."

The sheaths of insulating space are elasticated, and follow Hook's law, which also governs the stretching of spiral springs. This means that the more they are stretched the greater the tension they exert. On the scale of galaxies and clusters of galaxies, the tension within these tubes gives rise to addition forces, and we call this *dark matter*. At cosmological distances the tension becomes very much greater, and within the context of Einstein's general theory of relativity, it causes the expansion of the universe to accelerate—this is what we call *dark energy*. These tubes of force form a cosmic mesh, or universal grid, linking every part of the universe to every other part, and I believe it is this mesh, or grid, that is responsible for quantum effects—for dark matter and for dark energy.

An Urban Analogy

Scientists have always used analogy to explain difficult concepts. At the end of Chapter 8 I use an urban analogy to try to clarify my ideas on the various levels of reality: Modern life in a big city benefits from a number of subsystems. Above the ground we have pedestrians, cars, buses, and ordinary trains, and the interactions between these are controlled, in general, by traffic lights and pedestrian crossings. However, on occasion, some vehicles, such as ambulances, fire engines, and police cars, are permitted to ignore traffic lights in order to carry out their duties efficiently. On these occasions they also do not have to follow the normal speed limits imposed on other vehicles. This I call a*bove-ground reality*. In the wider universe, this is the type of reality that can be investigated using normal scientific equipment, such as, for example, microscopes and telescopes.

Taxi cabs, police cars, ambulances, and fire engines can communicate with each other, and with their headquarters, via radios, which use electromagnetic waves. We can all communicate with each other using mobile phones, which also use electromagnetic waves. This too is part of above-ground reality, but it is invisible above-ground reality. This is similar to magnetic fields in the universe: We cannot see these fields, but they can be detected using the methods described in Chapter 2.

Let us suppose that there are intelligent beings on a planet on which it is too difficult to make tunnels, because of the hardness of the crust. Further suppose that these beings have powerful optical telescopes and sensitive radio telescopes with which they are studying Earth. They may be puzzled by the sight of people disappearing into holes in the earth, and suddenly reappearing. They may also wonder where the water comes from that powers large fountains, and what powers the many electric lights that come on at night. This will be because they do not know of the underground systems that are part of most large cities. For them, this is an invisible world of which they know nothing.

Below the surface of a city we have underground reality: underground railway systems that allow faster transport; water pipes; electric, telephone, and data cables; and the sewage system. All these help to make life in a city possible and bearable—and would explain the features extraterrestrials might find difficult to understand, if they had no concept of another level of reality.

Dark matter and dark energy are so called because we cannot see these components of the universe. Some aspects of subatomic particles point to a level of reality that is not directly visible, although we are aware of its consequences. According to my plasma space theory of matter, at the beginning, there would have been a large number of neutrons around, which would have decayed in the first few minutes into protons, electrons, and neutrinos. The neutrinos would have become detached from the protons and electrons, but the electrons would still be attached to the protons via lines of force that splay out in ordinary space and braided ropes of lines of force encased in sheaths of insulating space.

In the very high temperatures that existed in the early universe, there would have been little likelihood of an electron pairing up with the proton that came from the same neutron as itself. This would only have happened when the universe had expanded considerably and the temperature had dropped a great deal. When hydrogen atoms did eventually form, the electrons and protons born from the same neutron would most likely pair up with other electrons and protons, but they would still be attached to each other by means of the braided lines of force encased in sheaths of insulating space. This means that there is an expanding grid of these braided lines of force linking different parts of matter together. Because they are encased in insulating space, we are not aware of the electric lines within, just as we are not aware of heavy electric currents passing through copper wires enclosed in tubes of insulating material.

This grid is a cosmic "underground" along which information can pass faster than the speed of light, which is the limiting speed for movement and transmission in ordinary space. This is similar, in a very broad sense, to the fact that the speed limit for underground trains is higher than that for above-ground traffic. On occasion it is possible, especially where the tube line is relatively close to the surface of Earth, to feel the vibrations of the trains at ground level. I believe this is similar to 'feeling' the vibrations of the cosmic-wide grid that manifests itself as the pilot waves first proposed by Louis de Broglie.

No analogy is exactly right, and the urban analogy is no exception: Although occasionally new branch lines are added to the underground network, it is for the most part fixed. On the other hand, the cosmic mesh expands as the matter in the universe expands, so it is a dynamically evolving entity. (The consequences of this evolution will be diescussed in Chapter 7, and the urban analogy will be further detailed in Chapter 8.)

Chapter Five

Evidence for the Unseen

The main purpose of this chapter is to discuss the evidence for dark matter and dark energy. To set the scene, I will look more generally at how astronomers and other scientists have tried, in the past, to explain the behavior of visible objects in terms of invisible forces and fields. William Berkson, in his book *Fields of Force*, has this to say:

> The idea that there is some hidden cause to the changing world of our experience is common to both science and most religions. Perhaps at the root of both is a sense of wonder or awe at nature. In the case of religions the explanation has been that there is a god or gods who have control over the phenomena we observe. We may form some idea about the supernatural powers by reasoning or reading certain writings supposed to be inspired by those powers. So the mystery in religion consists, in part, in the exact nature of God and his powers. Scientific explanation has been of a different sort. It originated with the Greeks and was later improved by the men of the Renaissance.

Also, in the introduction, Berkson said, "Faraday's field theory and its successors were a new solution to an old problem, a problem which goes back at least to the ancient Greeks. The problem is: How does one body act on another? Some of the more specific questions the theory helps to solve are: Why does one body push another along, without penetrating it? How does a magnet cause a piece of iron some distance away to move? And how can an electrified body cause dust near to it to move towards it?"

We will look at a few examples of unseen materials and forces that have been invoked to explain observed motions, starting with the ancient Greek explanation for planetary motion.

Aristotle's Universe

In this universe the Earth was fixed; the planets (which at that time included the sun and moon) and the stars moved around the Earth. He believed that the stars were all fixed to one large sphere, and that this sphere was concentric with the Earth. This sphere was often called the *sphere of the fixed stars*, not because it was fixed, but because the stars were fixed with respect to each other, so they appeared as if they were fixed to this outer sphere. The sphere of the fixed stars was, in fact, supposed to spin around the Earth in slightly less than one day (or 23 hours, 56 minutes of our time). The space between the sphere of the fixed stars and the Earth was filled with seven other crystalline spheres (they had to be clear so that the stars could be seen), and to each sphere was attached one of the planets, the sun, and the moon. The sphere touching the fixed stars was attached to Saturn, Jupiter was attached to the next slightly smaller sphere, then Mars, then the sun, then Venus, then Mercury, and finally the moon.

The motion of the sphere of the fixed stars was transmitted downward via friction between neighboring spheres. Thus, the sphere of Saturn moved least with respect to the background stars, and the sphere of the moon moved most, because the friction drive was least effective in this latter case, and the lunar sphere was in contact with a fixed Earth that further reduced the effectiveness of the drive.

FROM KEPLER TO NEWTON

In an earlier chapter we saw that Kepler used the anima motrix and the magnetic fields of the sun and planets to explain his laws of planetary motion. Both these agencies were invisible, but they did serve to give some quantitative explanation for the observed motions of the planets. Newton went a stage further when he introduced the more general laws of motion and the law of gravitation. This law of gravitation was a precise mathematical statement of how bodies affected each other: "Every particle in the universe attracts every other particle with a force directly proportional to the product of their masses and inversely proportional to the square of the distances between them."

However, Newton's law of gravitation depended on instantaneous action at a distance, thus each body attracted every other body with a force that reached across space without the intervention of a medium. Some people felt that Newton had introduced occult ideas into science. For several decades this law, when fully applied to the bodies of the solar system, supplied a complete explanation for planetary motion.

Kepler's laws of planetary motion told us that the planets went around the sun in elliptical orbits, with the sun at one of the foci of each ellipse. He also gave us the law telling us how the speeds of the planets in their orbits were related to their distances from the sun. However, in Kepler's scheme, each planet was treated as if it was moving under the influence of the sun alone; it did not allow for the interaction of one planet with another. In Newton's scheme, the planets could interact with each other, even though the sun was the most massive body in the solar system, and hence exerted the dominant control over the planets. If there was only one planet going around the sun, then its elliptical path would remain fixed in space, with respect to the much more distant stars. If a solar system had two planets, then, according to Newton's law of gravitation, there would be a mutual attraction between these two bodies, which would vary as they orbited at different speeds around the sun. The main consequence of this interaction would be that the major axis of each ellipse would now move very slowly around the sun, in addition to the motion of each planet along its elliptical orbit. This motion is referred to as the *precession of the major axis*. Before the discovery of Uranus, all six planets interacted with each other, and these interactions contributed to the precessions of the major axes of each planet.

The Discoveries of Uranus and Neptune

Uranus was the first planet to be discovered by a telescope. It was discovered by the astronomer William Herschel, working in Bath, in March, 1781. With one of his telescopes he noticed a body that seemed to change its position against the background of stars. At first he thought it was a comet, and reported it as such, but further observations showed that it had sharp edges, so it had to be a planet. Irregularities in the orbit of Uranus helped astronomers to discover the planet Neptune.

Soon after the discovery of Uranus, astronomers starting plotting its orbit. Other astronomers found that the planet had actually been seen on about 20 previous occasions, but it was always catalogued as a star, because it was very slow-moving, and the observers who saw it did not realize it was changing its position from one night to the next. Not long after this, it became clear that the orbit of the planet calculated using the newer observations was slightly different from the orbit calculated using the older observations. What was the reason for this?

The French astronomer Bouvard (1767–1843), who was at one time director of the Paris Observatory, was one of the first scientists to draw attention to the problems with the orbit of Uranus. Bouvard, having already published accurate tables for the orbits of Jupiter and Saturn in 1808, undertook to produce a corrected version of D'Alembert's (1717–1783) tables for Uranus. When he published his new tables of Uranus in 1821, he wrote, "I leave it to the future the task of discovering whether the difficulty of reconciling [the data] is connected with the ancient observations, or whether it depends on some foreign and unperceived cause which may have been acting upon the planet."

John Couch Adams (1819–1892), who was born in Laneast, in Cornwall, went to study mathematics at Cambridge, and on July 3, 1841, when he was still an undergraduate, he wrote, "Formed a design in the beginning of this week, of investigating, as soon as possible after taking my degree, the irregularities of the motion of Uranus, which are yet unaccounted for; in order to find out whether they may be attributed to the action of an undiscovered planet beyond it: and if possible thence to determine the elements of its orbit, etc., approximately, which would probably lead to its discovery."

Some other astronomers, including George Biddell Airy, who was the Astronomer Royal at this time, believed that the problem of Uranus's orbit arose from a breakdown of Newton's law of gravitation at great distances from the sun.

Meanwhile, in France, the director of the Paris Observatory, Arago (1786–1853), suggested to Urbain Le Verrier (1811–1877) that he address himself to the problem of the anomalous motion of Uranus, and he started work straightaway. At this time, no one was aware of what Adams was doing in England, and likewise Adams was not aware that anyone else was working on the problem.

Adams then made a more detailed study of the problem, and by September 1845 he had deduced an orbit for the perturbing planet, calculated its mass, and predicted its position for October 1, 1845. Le Verrier and Adams were both breaking completely new ground, and their individual successes were a landmark in the history of science, because never before had scientists attempted to work out the presence of another body from its effects on the orbit of a body that had already been observed.

Adams wanted to communicate his findings, in person, to Airy, but all his attempts to meet up with the Astronomer Royal failed. On October 21, 1845, Adams made two separate attempts to see Airy, but, failing to do so, he left a copy of his manuscript at Airy's home on the second occasion.

On November 5, Airy wrote to Adams asking if the "new planet" could explain the anomalies in the observed position of Uranus in the sky, as well as distance from the sun. This question was designed to try to distinguish between the "new planet" theory and the "failure of the inverse square law" theory. Adams, however, did not reply, possibly because he saw Airy's reluctance to meet him as a snub.

Le Verrier published two papers on the possibility of a "new planet"; one on November 10, 1845, and another on June 1, 1946. Airy received news of Le Verrier's work on June 23, and saw immediately that the predictions were almost identical to those of Adams. Airy asked Le Verrier the same question he had asked of Adams, and received the reply that the deviations in the orbit of Uranus were due to the "new planet."

Airy made no attempt to search for the planet, but asked Professor Challis to begin a search at the Cambridge Observatory. Reluctantly Challis began his search on July 29, 1846, and continued his observations into August, noting the observable stars that could be seen in the search area suggested by Le Verrier and Adams, but unfortunately he did not compare the observations he made on the different dates. Subsequent perusal of his records, sometime later, after the planet had been discovered in Germany, showed that Challis had recorded the positions of the "new planet" on two occasions; once on August 4, and again on August 12.

John Herschel, son of William Herschel, addressed a meeting of the British Association for the Advancement of Science in Southampton on September 10, 1846, when he spoke of the "new planet," saying, "We see it as clearly as Columbus saw America from the shores of Spain. Its movements have been felt, trembling along the far-reaching line of our analysis, with a certainty hardly inferior to that of ocular demonstration."

Le Verrier wrote to the German astronomer Galle, on September 18, asking him to search for the planet at the Royal Observatory in Berlin. Galle wrote to Le Verrier on September 25, saying: "Monsieur, the planet of which you indicated the position really exists."

This then was the story of the discovery of Venus, which was a major triumph for mathematical astronomy. All three men—Adams, Le Verrier, and Galle—are rightly credited with its discovery.

Anomalies in the Orbit of Mercury

Not long after the discovery of Neptune, astronomers began to notice irregularities in the orbit of Mercury, which could not be explained in terms of the perturbations of the other planets on its orbit. Earlier on in this chapter I drew attention to the fact that, if there was only one planet in the solar system, then the major axis of its elliptical orbit would remain fixed with respect to the distant stars. The point on the elliptical orbit of a planet at which it is closest to the sun is called its *perihelion*, and the point at which it is furthest from the sun is called its *aphelion*. Giving the orientation of the orbit in space is normally done by specifying the position of the perihelion, so any changes in its orientation is normally given in terms of the shifting of the perihelion. The main problem with Mercury was that its perihelion was shifting more than one would expect on the basis of the perturbations of all the other planets of the solar system, so the problem was usually referred to as the *perihelion excess of Mercury.* Although perihelion excesses also existed for some of the other planets, it was more noticeable in the case of Mercury. This was because of its closeness to the sun, making its orbital period shorter than those of the other planets, and hence its perihelion excess showed up more quickly than it did for the other planets.

A few theories were put forward to account for this excess. One theory was that there was yet another planet between Mercury and the sun. Another was that a ring of dust existed between the planet and the sun. Yet another proposed that the sun was not quite spherical, but was flattened at the poles, rather like our own Earth. Although searches were made to confirm these theories, no evidence was found to support any of them, so they were abandoned.

By far the most successful class of models came from mathematical modifications of Newton's law of gravitation. Some scientists suggested that perhaps this law was not simply an inverse square law, but one in which the power to which the distance was raised was slightly different from two. This did not meet with observational confirmation, so other approaches were tried. One that gave results closer to the observed perihelion access was based on the analogy between electricity and magnetism. It was known that two negatively charged particles would repel each other according to Coulomb's law. However, it was also known that if the two charges were moving in the same direction they would constitute two parallel electric currents, resulting from their motion. Such currents would give rise to magnetic fields, and because the currents were in the same direction, they would tend to attract each other. The force of attraction would not be sufficient to overcome their electrostatic repulsion, but it would weaken it. It was reasoned that Newton's law was similar to an electrostatic attraction, so was it not possible that a planet in motion could give rise to an additional force akin to the magnetic force in the case of moving charged particles. Some people called this a *gravo-magnetic force*, and its application to the orbit of Mercury accounted for some of the observed perihelion access. By far the most successful explanation was provided by Einstein's general theory of relativity, which was essentially a new theory of gravitation.

EINSTEIN'S GENERAL THEORY OF RELATIVITY

In 1905, Einstein formulated his special theory of relativity, which showed how the laws of classical physics had to be modified if one was comparing the measurements of mass, length, and time made by two observers moving at different, constant speeds in straight lines with respect to each other. A cornerstone of this theory was the speed of light. His theory showed that the speed of light was not only the ultimate speed limit of all matter and information, but also that every observer would measure exactly the same speed, no matter how fast he or she was moving at any speed less than that of light. He also derived the most famous mathematical equation in the history of science: $E = mc^2$, which showed that if a small amount of matter was destroyed, a large amount of energy would be released. This tells us why our sun and most of the stars are able to emit enormous amounts of energy for millions of years. It also tells us why the atomic bomb and the hydrogen bomb have such destructive powers. However, gravitation was not part of the theory of special relativity, so Einstein started on the quest for a more embracing theory that would include gravitation. This was to become the general theory of relativity.

Soon after the special theory of relativity was further developed by Einstein and others, it became apparent that it achieved its most elegant form of expression in four-dimensional space, with three dimensions of space, and one of time. This approach was to become even more important in the general theory of relativity. Essentially, Einstein showed that space and time were curved near massive bodies, and that in this curved space-time, bodies that were not acted on by other forces would follow a special path, called a *geodesic.* Thus, space-time in the neighborhood of the sun would be curved, and all the planets would follow geodesic paths around the sun.

When these concepts were applied to the orbit of Mercury, they provided a full explanation for the perihelion excess for this planet, and also for other planets close to the sun. Was there any other observations that could be undertaken to test the theory of curved space-time in the neighborhood of the sun?

Einstein showed that if a ray of light from a star passed very close to the sun, then it would be bent. This meant that it should be possible to see a few stars that were actually *behind* the sun, if it were possible to blot out the light from the sun itself. During a total eclipse of the sun, the moon does just that! Professor Arthur Eddington, from Cambridge University, and Sir Frank Dyson, who was Astronomer Royal at Greenwich at the time, decided to organize two expeditions to observe the eclipse of May 29, 1919, to test this aspect of the general theory. One expedition, organized by Dyson at Greenwich, went to Sobral in Brazil, and the other, organized by Eddington at Cambridge, went to the Island of Principe off the west coast of Africa.

Einstein had carried out two different calculations on the bending of light by the sun. In the first (published in 1911) he assumed that because light behaved as if it consisted of particles (called *photons*), and because particles had energy, they must also have mass. This meant that a massive body such as the sun could attract the photons in a ray of light, as it passed close to the sun, according to Newton's law of gravitation. The other calculation (published in 1916) made use of the curvature of space-time, which was predicted by the general theory. The bending of light on the basis of the second calculation was twice that given by the first calculation.

The paper announcing the results of the two eclipse expeditions, by Dyson, Eddington, and Davidson, had this to say: "Thus the results of the expeditions...can leave little doubt that a deflection of light takes place in the neighborhood of the sun, and that it is of the amount demanded by Einstein's generalized theory of relativity...."

The philosopher Alfred North Whitehead was present at the meeting when this paper was read, and he recalls, in *Science and the Modern World*:

> It was my good fortune to be present at the meeting of the Royal Society of London when the Astronomer Royal for England announced that the photographic plates of the famous eclipse, as organized by his colleagues in Greenwich Observatory, had verified the predictions of Einstein, that rays of light are bent as they pass in the neighborhood of the sun. The whole atmosphere of tense interest was exactly that of the Greek drama: We were the chorus commenting on the degree of destiny as disclosed in the development of a supreme incident.

The paper, more than any of those by Einstein himself, brought Einstein's work to the general public, and made him famous overnight.

This aspect of the general theory is called *gravitational lensing*, because the gravitational field is acting, to some extent at least, as a lens made of glass. It has been used to great effect to provide evidence for dark matter in the universe.

Evidence for Dark Matter

There are four main areas of research I will consider, and one indirect method. The first area of research is the rotation curves of galaxies—these are the ways in which astronomers measure the speeds of stars at different distances from the centers of galaxies. This information can, in turn, be used to work out the distribution of mass within a galaxy. Such studies reveal that there are still considerable amounts of matter beyond the visible edges of galaxies.

The second is the behavior of galaxies in large clusters of galaxies. It was initially believed that a cluster of galaxies was kept together by the mutual gravitational attraction between the individual members of the cluster. However, it was eventually shown that there was not enough mass in all the visible galaxies to account for the cohesion of the cluster, so dark matter had to be invoked. In recent years it has been found that the light from galaxies behind large clusters can be bent by the total amount of material in the cluster, in much the same way as it is possible to see stars behind the sun during a total eclipse. This gravitational lensing effect has provided more evidence for the existence of dark matter.

Third, once X-ray observations of the universe became possible, it was revealed that there were considerable amounts of hot gases in clusters of galaxies. The fact that this high-temperature gas did not disperse meant that there had to be an unseen source of extra mass within the clusters.

The indirect method involves the large-scale motions of galaxies toward superclusters of galaxies.

Evidence From the Rotation Curves of Galaxies

In some respects, there are similarities between our solar system and our galaxy. The planets, asteroids (sometimes referred to as minor planets), and comets orbit the sun, and although the bodies interact with each other, they are dominantly under the gravitational attraction of the sun. The planets and asteroids lie very much in the same plane, and they have orbits that do not differ all that much from being circular, but the comets have much more elliptical orbits, and they are not confined to the plane of the planets and asteroids. The speeds with which bodies orbit the sun depends on their distances from the sun, but the general law they follow is that the speed with which a given planet orbits the sun is inversely proportional to the square root of its distance from the sun. Our Earth, for instance, has an average speed of about 66,000 miles per hour, but Saturn, which is about nine times further away from the sun, has an average speed of about 22,000 miles per hour.

Our galaxy has a central bulge, called the nucleus, which could be seen as somewhat similar to the sun, and the majority of bright young stars orbit the nucleus in nearly circular orbits in a disc that is thin compared to its diameter. This means that the young stars have some similarities with the planets and the asteroids. The older stars and the globular clusters in the halo have much more elliptical orbits, and they are not confined to the disc of the galaxy. This means that they are in some ways similar to the comets of the solar system. There the similarities end. Only a proportion of the mass of the galaxy is concentrated in the central bulge, and the stars in the disc, and those in the halo, all make some contribution to the gravitational field in which they move. This means that the speeds with which stars orbit the central bulge depend on their distances from the center in a different way from the speed-distance relationship for the solar system.

The speed-distance relationship for a galaxy is called its *rotation curve.* Near the visible edges of a galaxy one would expect the speed-distance relationship to approach that of our solar system—the speed would very nearly be proportional to the inverse of the square root of distance from the center. However, this is not the case; the speed does not vary very much with distance from the center, and astronomers say the rotation curve is flat. This implies that there is dark matter beyond the visible edges of galaxies. The indications are that the total mass is increasing at a rate proportional to distance from the center. It is generally believed that the dark matter is in the form of an invisible, roughly spherical halo, which has a radius of about

3 to 4 times larger than the radius of the visible disc. This is very convincing evidence for the existence of dark matter in galaxies.

Evidence From Large Clusters of Galaxies

Before discussing the nature of this evidence, I will make a brief excursion into planetary atmospheres, because I wish to use this as an analogy. Some planets have atmospheres, and others do not—what are the determining factors? The most important are the strength of the force of gravitation at the surface of the planet, and the speeds of the atoms and molecules of the atmosphere at the average temperature of the gases that make up the atmosphere. The surface gravity depends on the total mass of the planet and its radius. Increasing the mass increases the surface gravity, but the gravitational tug of the planet is weakened by how far one is away from the center of the planet. For example, the moon has a radius of about a quarter of that of Earth, and a mass of 1/80 of the Earth. If one squares the radius and divides it into the mass of the moon, it gives a surface gravity of about 1/5 that of Earth.

Escape velocity is the velocity that a body has to have if it is to totally escape the gravitational tug of a planet. The temperature of a gas is the measure of the average speed with which its atoms or molecules are moving. At a given temperature the molecules of a light element will be traveling much faster than those of a heavier element. So whether a given element will remain in the atmosphere of a planet will depend on how the average speeds of its molecules will compare with the escape velocity of the planet. The Earth's atmosphere has no hydrogen, but it does have oxygen and nitrogen. This is because at the temperature of our atmosphere, the hydrogen molecules will have speeds that exceed that of the escape velocity, so the hydrogen would have escaped to space, and the lower speeds of the oxygen and hydrogen would mean that they remain part of out atmosphere. The moon has no atmosphere because all the molecules that make up our own atmosphere would have been traveling at speeds that exceeded the escape velocity of the moon.

The galaxies in a large cluster of galaxies can, to a first approximation, be seen as molecules moving in the total gravitational field of the cluster. Their average speeds can be seen at the temperature of the cluster. Using the Doppler effect, one can work out the average speeds of the galaxies.

If one uses the total mass of all the visible galaxies in the cluster, and their average distances apart, to work out the "surface gravity" of the cluster, one can then calculate the escape velocity of the cluster. It turns out that the average speeds of the galaxies are greater than that of the escape velocity. This implies that there must be an unseen component of the cluster adding to its total mass. This is very similar to the method used by Fritz Zwicky (1898–1974) in 1933 to deduce the presence of dark matter in the Coma Cluster of galaxies.

Evidence From Gravitational Lensing

In recent months, two groups of astronomers have used gravitational lensing to provide convincing evidence for the existence of dark matter. Douglas Clowe, of Ohio University, and Dennis Zaritsky, from the University of Arizona recently disussed the strange case of the Bullet Cluster of Galaxies, in their article "Shot in the Dark" in the February issue of *Physics World*. They started by pointing out that in 1933 the Swiss-American astronomer Fritz Zwicky had found that the galaxies in the Coma Cluster (known also as the Bullet Cluster), were moving so fast with respect to each other that they should have escaped from the cluster by the time he made his observations. They then went on to point out that the stars in this cluster, which had been observed by Zwicky and others, were not its main component. Studies of the cluster by X-ray astronomers seemed to show that about 90 percent of the visible matter was in the form of a diffuse and very hot ionized gas. Using all the available information they concluded that, "...the Bullet Cluster has to have at least twice as much mass in dark form than in a visible baryonic form."

In order to perform gravitational lensing measurements of the Bullet Cluster, Clowe, Zaritsky, and their coworkers obtained deep optical images using the European Southern Observatory's Wide Field Imager, the Very Large Telescope in Chile, the Magellan Observatory, and the Hubble Space Telescope. Another item in that issue of *Physics World* was a brief report in the section on "Frontiers," which reported on a large map of the distribution of dark matter in the universe, which astronomers had constructed using observations from the Hubble Space Telescope: "The 3D map...reveals a network of dark-matter filaments that coincides with the distribution of normal matter such as galaxies. As such, it provides strong evidence that the universe owes it structure to the gravitational pull of dark matter."

These astronomers had also used gravitation lensing observations to arrive at their conclusion.

Hot Gas in Galaxies and Clusters

One of the early surprises of X-ray astronomy, once it became technologically possible, was finding that clusters of galaxies were intense X-ray sources. The X-rays did not come from the galaxies themselves, but from very hot and rarefied gases between the galaxies. The temperature of this gas was typically about 10,000,000 degrees Celsius. To return to our planetary atmosphere analogy, this "cluster atmosphere" can only remain close to the cluster if the gravitational field is strong enough. Calculations show that there is not enough material in the visible components of the cluster to retain this atmosphere, so there has to be an unseen, or "dark matter" component.

The Great Attractor

A group of astronomers, known by some as the "Seven Samurai" (Burstein, Davies, Dressler, Faber, Lynden-Bell, Terlevich, and Wegner) made an important discovery concerning the distribution of galaxies in space. In the 1980s they found that, not only were galaxies unevenly distributed in space, but also galactic superclusters were separated by incredibly huge voids of visible ordinary matter. The Great Attractor is one such structure. Apparently the Great Attractor is pulling in millions of galaxies in a region of the universe that includes the Milky Way, the surrounding local group of 15 to 16 nearby galaxies, the larger Virgo Supercluster, and the nearby Hydra-Centaurus Supercluster. Based on the observed galactic velocities, the unseen mass within the voids between the galaxies and clusters of galaxies is estimated to total around 10 times more than the visible matter in this region of the cosmos, and so it must be composed mostly of dark matter.

Evidence From Cosmology

Cosmology is the area of research concerned with the large-scale structure and evolution of the universe. It really started with Einstein's general theory of relativity. According to this theory the large-scale "shape" of space can be flat or curved, depending on the average density of the matter within the universe. An important question cosmology has to answer is concerned with the ultimate fate of the universe. It is generally accepted that the universe started with the big bang, and that it has been expanding ever since. If the average density of matter is equal to, or less than, a special value, called the *critical density*, then space is flat and the universe will expand forever. However, if the average density of the universe is greater than the critical density, then space will be curved, the universe will continue to expand for a while, then slow down, the expansion will stop, and the universe will start to contract. Observational cosmology seems to indicate that the universe is flat, and hence the density is at least equal to the critical density.

Evidence that the universe did indeed start with the big bang comes from the cosmic microwave background radiation. The initial fireball would have been very hot, but as the universe expanded, the radiation would have cooled down, so today it is just 3 degrees above absolute zero. This radiation has been detected by astronomers, and except for very small fluctuations, it is remarkably uniform. Because the matter in the universe is separated by large distances, how is it that the cosmic microwave radiation is so uniform? This uniformity must have arisen when the radiation and the matter occupied a much smaller volume of space. However, in order to have maintained this uniformity, the early universe must have undergone an extremely rapid period of expansion. This period is now called the *inflationary period*, and it was first proposed by Alan Guth (born 1947), in 1979.

According to Jonathan Sherwood, in a *Cornell Chronicle* article entitled "Lecturer describes what may have happened just before the big bang," Professor Guth, Victor F. Weisskopf professor of physics at MIT, explained that the classical big bang was not really a bang; it merely described what happened just after this singular event. However, the concept of inflation, first introduced into cosmology by Guth himself, could explain the big bang. Inflation arose from the application of some of the concepts of modern particle physics into Einstein's general theory of relativity. When this was done it showed that in the early moments of the universe, gravity could become repulsive rather than attractive.

If inflation is right, then, because luminous stars and galaxies only contribute 0.5 percent of the density needed for it to be right, 99 percent of the universe must be in the form of dark matter. All in all, the evidence for dark matter is convincing, but what is the make-up of this matter?

THE CANDIDATES FOR DARK MATTER

These candidates fall into two groups. The first group is called MACHOs, which stands for Massive Astrophysical Compact Halo Objects, and the second group is called WIMPs, which denotes Weakly Interacting Massive Particles. The first group is made up of ordinary, or baryonic matter, but the latter group is made up of non-baryonic matter. Astronomers look for the first group, and particle physicists look for the second group. Let us look at each group separately, in slightly more detail.

MACHOs

There are two types of objects in this group: brown dwarfs and black holes.

Brown dwarfs are stars that did not quite make the grade: Stars are born out of the gas in the interstellar medium, when clouds collapse under the force of gravitation. Stars that have similar masses to our own sun will have enough matter in them to gravitationally squeeze their interiors so hard that their internal temperatures will increase considerably, to the point that the thermal collisions between nuclei will initiate nuclear reactions, mainly turning hydrogen nuclei into helium nuclei. When this happens, the protostar becomes a real star. A brown dwarf will have heated up to some extent, as it collapsed from a gas cloud, but the temperatures will not be high enough to initiate nuclear reactions in its interior, so it will only be able to radiate the heat it acquired from the collapse. This means that it has very low luminosity, and hence it will be difficult to detect in the halo of our galaxy.

The Hubble Space Telescope has been used to look for brown dwarfs, but the results have been disappointing, leading astronomers to estimate that they only constitute about 6 percent of galactic halo matter.

Black holes are the end product of a very massive star that has used up all its nuclear fuels, so there is no internal energy to provide the heat and the pressure to support its further collapse under gravity. The remaining mass in the star is so great that degeneracy electron pressure cannot support it to become a white dwarf, and degeneracy neutron pressure is not sufficient to support it to become a neutron star. This means that it just continues to contract under the force of gravity. The surface gravity of the black hole gives rise to an escape velocity greater than the speed of light, so it swallows up its own light, and this is why it is invisible (hence the name *black hole*). However, its presence can be deduced from its effects on its surroundings. It can cause gravitational bending of light from stars that lie beyond it, and if it has a companion star, such a star will orbit the unseen black hole, and we can deduce the mass of the black hole by measuring the orbital velocity of its companion, if we also know how far apart they are.

WIMPs

WIMPs could provide another possible solution to the problem of dark matter. These particles interact through the weak nuclear force and through the force of gravity. They cannot be seen directly, because they do not interact with electromagnetism. They do not interact strongly with atomic nuclei because they do not interact with the strong nuclear force.

WIMPs are even more difficult to detect than MACHOs, because they interact so weakly with matter. However, as a WIMP passes through matter, it is just possible that it will interact occasionally with a particle of ordinary matter, and we can detect the outcome of such an interaction. One such experiment uses a very large crystal cooled to almost absolute zero, which is surrounded by sensitive instruments that can detect slight changes in the heat of the crystal. Another experiment is under way in Antarctica, called AMANDA—Antarctica Muon and Neutrino Detector Array. The "crystal" in this case is the Antarctic ice sheet itself, so the instruments are placed deep with in the ice. Neither experiment has yield any results to date, so more work is necessary before any positive conclusions can be reached.

EVIDENCE FOR ACCELERATING EXPANSION AND DARK ENERGY

The most convincing early evidence for an acceleration in the expansion of the universe came from a type of exploding star known as Type 1a supernovae. These are carbon- and oxygen-rich white dwarfs that undergo thermonuclear explosions. These stars are 40 percent more massive than the sun, but they have a radius 100 times smaller. In an earlier chapter we saw that there was a limit to the mass of a white dwarf, called the *Chandrasekhar limit*, which is 1.4 times the mass of the sun. However, in a binary system the strong gravitational field of the white dwarf can pull matter off the companion star until it exceeds the 1.4 solar mass limit, and the excess mass destabilizes the star, which then explodes as a supernova. The luminosities of these supernovae are all very similar, so one can use these stars as "standard candles." That means that by comparing their average luminosities with the luminosities we can measure on the surface of Earth, we can work out the distances to these objects.

Two teams of astronomers, the High-Z Supernova Search Team and the Supernova Cosmology Project, in the mid-1990s, initiated observational programs to measure the distances and speeds of Type 1a supernovae. The results obtained from about 100 supernovae were astonishing! They discovered that the fastest-moving supernovae were fainter, and hence further away, than they should have been in a universe whose expansion rate was slowing down. In other words, the evidence was consistent with a universe whose expansion rate was actually increasing!

This result was so surprising that some astronomers began to question the basis of the method. How can we be certain that the light from these supernovae is indeed diluted and dimmed due to traveling the greater distances that result from the accelerated expansion of the universe? It should be noted that Type 1a supernovae are not quite standard candles, but in recent years their luminosities have been standardized. Detailed observations of nearby supernovae at known distances have revealed a pattern that can be used to calibrate their luminosities using their light curves and spectra. The results so obtained have confirmed the accelerated expansion of the universe. So how do we explain this acceleration? The simple answer is negative pressure, so let us explore this concept.

Negative Pressure

Writing in *Physics World* in May 2004, in an article called "Dark Energy," Robert Caldwell, from Dartmouth College, had this to say: "The biggest mystery of the cosmic acceleration is not that it suggests that two-thirds of the universe is made of stuff that we cannot see, but that its suggests the existence of a substance that is gravitationally repulsive."

Negative pressure is also mentioned by Lee Smolin in his book *The Trouble with Physics*, published in 2006:

> Most kinds of matter are under pressure, but the dark energy is under tension—that is, it pulls things together rather than pushes them apart. For this reason, tension is sometimes called negative pressure. In spite of the fact that the dark energy is under tension, it causes the universe to expand faster. If you are confused by this, I sympathize. One would think that a gas with negative pressure would act like a rubber band connecting the galaxies and slow the expansion down. But it turns out that when the negative pressure is negative enough, in general relativity it has the opposite effect. It causes the expansion of the universe to accelerate.

The next question we need to consider is, what are the possible candidates for negative energy?

Possible Candidates for Dark Energy

This section is based on an article on dark energy, written by Robert Caldwell, from the Department of Physics and Astronomy at Dartmouth College, which appeared in *Physics World* in May 2004. Under the heading "Dark Energy: The Suspects," he listed the following four candidates.

The Cosmological Constant

Albert Einstein introduced this term to explain the fact that, at the time he formulated the general theory of relativity, there was no evidence for the expansion of the universe. At very large distances between galaxies, this term gave rise to a repulsion that would counteract the usual gravitational attraction, and this would contribute to the stabilizing of the universe. The discovery of the expansion of the universe, by Edwin Hubble, made it unnecessary to introduce this term into models of the large-scale structure and evolution of the cosmos. However, the discovery that the expansion of the universe was, in fact, accelerating, has led some cosmologists to reintroduce this term.

Quintessence

This is a form of energy with negative pressure. However, unlike the cosmological constant, it is dynamic and varies with space, and its energy, density, and pressure slowly decays with time.

Other Forms of Vacuum Energy

In one model, quantum effects of a quintessence-like field lead to modifications of general relativity, whereas other models suggest that the dark energy density actually grows with time, causing the universe to end in a catastrophic "big rip."

Modifications of General Relativity

Various attempts have been made to modify Einstein's general theory of relativity. Unfortunately, many predict violations of one of the basic principles of the theory—in particular, the principle of equivalence. This principle postulates that if an experimenter was somewhere in space, in a lift with no windows, and found him- or herself pulled to the floor of the lift, they would be unable to decide if this was due to a gravitational field from a body outside and below the floor of the lift, or whether the lift was itself accelerating upward.

In Chapter 7 we will consider another candidate that can explain the accelerating expansion of the universe.

Chapter Six

Relativity and Quantum Theory

The first 30 years of the 20th century saw the emergence of two new theories in physics that completely overturned the basic ideas of what was to become known as classical physics, which consisted mainly of Newtonian mechanics and electromagnetic theory. The two theories were quantum mechanics and relativity. More precisely we should refer to them as two *classes* of theories, as they both consist of two related but distinct parts. Quantum theory really consists of nonrelativistic wave mechanics and relativistic quantum mechanics. Relativity consists of special relativity and general relativity. In this chapter I will discuss these classes of theories and highlight the important fact that there is a conflict between the two at a very deep level.

The Need for a Theory of Relativity

The physicists of the 19th century believed that electromagnetic radiation propagated through a medium called the *ether*, resembling, in very general terms, the propagation of sound waves through air. Between 1881 and 1887 the American physicist Albert Michelson (1852–1931) built and improved on a sensitive optical device, called an interferometer, to try to measure the speed of the Earth through the ether. All his attempts to do so ended with the same result, including the final one, in which he was joined by Edward Morley (1838–1923): the Earth was at rest in the ether! It was such an astonishing result that we really need to look at the experiment in more detail.

The idea of this experiment can be understood by considering the progress of two swimmers in a river. The first person swims across the river (all the water across the width is moving at the same speed) and back. Because he or she wishes to cross the river at right angles, he or she can only do so by heading upstream in each case, so as to allow for the fact that the river will carry him or her downstream. The other person, who swims at exactly the same speed as the first, swims the same total distance as the first, but he or she goes upstream and then downstream, rather than across the river. If one works out the times it takes the two swimmers to complete their round trips, it turns out that they are different. The Michelson-Morley experiment used this principle to try to measure the speed of the Earth through the ether.

The two "swimmers" were beams of light, from a single source, split into two beams of identical path lengths. The "river" was the ether, which, from the point of view of a piece of apparatus attached to the Earth, would seem to flow past the equipment. One beam traveled parallel to the direction of the Earth in its orbit, and the other perpendicular to this direction. The light from the two beams were combined to produce an interference pattern. Michelson and Morley reasoned that there should be a time delay between the two beams, which would be different for two different orientations of the interferometer—one position being the normal, and one with the interferometer rotated through 90 degrees. The difference could be detected, they assumed, by a shift in the interference pattern produced by combining the two beams. They found no such shift, which seemed, at first, to indicate that the Earth was at rest with respect to the ether.

In 1892 the Dutch physicist Hendrik Lorentz (1853–1928) offered another explanation for the negative result of the Michelson-Morley experiment. He assumed that the forces holding molecules together were electromagnetic in origin, and they would therefore propagate through the ether. If this was the case, then the particles of which the body was composed, moving through the ether, would be pushed together parallel to the direction of motion. Hence, the arm of the interferometer parallel to the Earth's motion shortens slightly with respect to the other. Lorentz was able to show that the amount of shortening was precisely that needed to cancel out the effect of motion through the ether expected from Maxwell's theory of electromagnetism. An Irish physicist, George Fitzgerald (1851–1901), put forward a similar explanation to explain the failure of the experiment to detect the Earth's motion. Although it is not clear if Einstein was aware of the experiment, or these explanations, his special theory of relativity was able to explain the result obtained by Michelson and Morley. Einstein was able to demonstrate that the measurements of length and time by observers traveling at different speeds, and in different directions, depended on their relative speeds and directions. This meant that the distance traveled by a light beam in the direction of the Earth's motion would appear to be different, to an observer on the Earth, from the distance traveled by a beam of light at right angles to the Earth's motion. Besides its other consequences, Einstein's work showed that there was no need to postulate the existence of the mysterious ether.

The Special Theory of Relativity

The laws that govern our day-to-day lives—our interactions with other people, our births, our marriages, and our deaths—will, of course, vary, depending on the country in which we are living, because they are human laws, constructed by people for the smooth running of societies with given sets of values. The laws of physics are rather different in character. They should not depend on where they are discovered, or on the locations in which they are applied and tested. They are required to be valid throughout the observable universe.

The concept of the universality of the laws of physics is relatively recent in the history of science, and can be traced back to Newton's formulation of the laws of motion and the law of gravitation. In Aristotle's cosmology, the laws of physics were not universal. There were two distinct sets of laws in his universe: one for the superlunar sphere, the region beyond the moon, and one for the sublunar sphere, the region below the moon. Kepler's laws of motion were the traffic rules for the planets—they did not apply to the motion of objects on the surface of Earth. Newton made an important discovery concerning the character of physical laws—they must be valid no matter where they applied in the universe.

Einstein discovered another characteristic of the laws of physics, which he used as a basic concept of the special theory of relativity: They had to be true for all observers that were moving at constant speeds in straight lines with respect to each other. In other words, the mathematical formulae used to describe the laws of physics had to be independent of the motion of the observer—provided they were traveling at constant speeds in straight lines. According to Einstein, if one were to test these laws in a laboratory at rest on the surface of Earth, in a moving train, in a spacecraft moving along in a straight line at constant speed, or at the bottom of a mine, one should always find these laws to be true.

The other basic principle of Einstein's special theory of relativity was that the speed of light was not only the same throughout the universe, but it also did not depend on the speed of the observer. The application of these basic principles to the modification of the laws of classical physics had many far-reaching consequences. The special theory of relativity required that the measurement of a length would be different for observers moving with different constant speeds in straight lines with respect to each other. Thus, for example, the length of a spacecraft as measured by an astronaut working outside the spacecraft would be different from its length as measured by a telescope fixed to the Earth.

Another consequence of special relativity was concerned with the masses of particles. One can find the mass of an electron by the extent to which it is deflected in a magnetic field of known strength. If one measured the mass of an electron moving close to the speed of light, then it would be much more massive than an electron moving at a much slower speed. This result of the special theory has been verified over and over again in the enormous machines used to smash subatomic particles together. Another consequence of the theory is that, from the point of view of an observer at rest, the clocks onboard a spacecraft traveling close to the speed of light would seem to have slowed, whereas no change would be detected by astronauts onboard the craft. All these consequences of the special theory of relativity are particularly important for subatomic physicists working on fast-moving particles, and for astronomers making observations on distant galaxies that are receding from us at very high speeds.

Mass, Energy, and the Speed of Light

We have already seen that the special theory requires that the mass of a particle will increase as it gets close to the speed of light. The equations of relativity tells us that at the speed of light a particle will have infinite mass, so this means that there is not enough energy in the universe to give a particle this speed, and hence no particle can travel faster than light. The theory also leads to the conclusion that if a small amount of mass is destroyed, then it will be manifested as energy. It also tells us that any localization of energy has mass.

When four hydrogen atoms are forced, by the high temperatures within the stars, to combine with each other to form one helium atom, then there is some loss of mass. It is this loss of mass that manifests itself as energy, and it is this energy that provides the fuel for most ordinary stars. The transmission of information normally requires the transfer of energy, and because energy has mass, no information can be transmitted faster than the speed of light. This means that the speed of light is the speed limit for matter and information in normal space.

Relativity and the Michelson-Morley Experiment

The special theory of relativity explains the negative result obtained by the Michelson-Morley experiment. Consider two measuring rods of equal length when they are at rest next to each other. Now consider that the one is at right angles to the other, and that one of them has its length in the direction of motion of the interferometer. The length of the one in the direction of motion will appear to be shortened with respect to the other. Thus, the theory, by clarifying our concepts of measurement, is able to explain why there is a difference in the measurements of the lengths the two beams in the Michelson-Morley experiment have to traverse, without requiring the physical change of length suggested by Lorentz and Fitzgerald.

World Lines in Space and Time

The special theory of relativity is most eloquently represented in the space-time continuum, which consists of three dimensions of space and one of time. It is virtually impossible to imagine four-dimensional space, but we can use analogies to illustrate the basic ideas—in many cases it is not always necessary to consider all three dimensions of space.

The position of a ship can be given in terms of its latitude and longitude at a given time. Imagine plotting its position on a chart drawn on a transparent sheet of Perspex at a given time, and its position at a later time on another such sheet. Consider plotting its position every hour on a series of sheets, and then stacking them on top of each other, in sequence, separating each sheet from the next by means of Perspex blocks placed at the corners. If instead of just plotting the positions at subsequent times, we had actually drilled a small hole in each sheet, then threaded a bit of cotton through each hole, the thread would represent the progress of the ship over the surface of the Earth, as well as its progress through time. This thread represents the "world line" of the ship through space-time. Even an object at rest has a world line, but in this case the world line would be moving only through time—it would be a vertical line through our sheets of Perspex.

The orbit of a planet around the sun is really an ellipse, but for most planets the orbits are more like circles to a reasonable degree of accuracy. However, the world line of a planet in space-time would be like a spiral, because the planet is not only orbiting the sun, but also progressing in time. Each and every particle in the universe has a world line in space-time, but the shapes of these world lines vary a great deal, depending on how the particle moves in space. The world lines of two particles that have collided with each other will intersect at the point of space and at the time at which they collided. Particles that have been at the same point in space, but at different times, will not intersect. In order to catch a bus, your world line has to intersect with that of the bus, at the bus stop, at the time the bus is due to arrive at the stop.

The General Theory of Relativity

Newton's law of gravitation did not fit into Einstein's special theory of relativity. One reason for this is that it requires instantaneous action at a distance, and this is in conflict with the requirement of special relativity that no information can be transmitted at a speed faster than that of light. Einstein's attempts to formulate a new theory led him to the general theory of relativity, which is really a theory of gravitation.

There are two basic principles to the general theory. The first is a statement concerning, once again, the character of the laws of physics. The theory requires that the laws should be stated in such a way that they do not depend on where the laws are applied, and they do not depend on how the observer is moving. This means that its requirements are more general than those of the special theory (which require the laws to be true for observers traveling at constant speeds in straight lines with respect to each other). The results of the special theory are strictly not valid when one is close to a strong gravitational field, or if one is changing one's direction of motion, or if one is changing the speed at which one is moving. The results of the special theory have to be replaced by those of the general theory in these cases.

The second basic principle is called the *principle of equivalence.* This principle asserts that gravitation and acceleration have a great deal in common, and that they are, in a sense, equivalent. We have all had some experience of this assertion: When a lift (or elevator) first starts to accelerate upward, we feel a slight increase in our weight, and a slight sinking feeling in our stomachs. When the lift stops we feel a slight feeling of weightlessness. This means that we can simulate an increase in the force of gravitation on our bodies by accelerating upward, and we can simulate a decrease in the force of gravitation on our bodies by accelerating downward. We can also experience gravitation-like forces when we are sitting in a car that is accelerating rapidly. In this case we are pressed back against the seat by the resulting force. Astronauts are most aware of these G-forces when they are accelerating upward in their spacecraft. In order to learn to cope with these forces, they are trained in a centrifuge.

World Lines in General Relativity

The two basic ideas just discussed lead to several conclusions of far-reaching importance. The first concerns the motion of particles. According to Newton's first law of motion, a particle will continue in its state of rest, or its motion in a straight line at a constant speed, unless it is acted on by a force that will tend to change either of the two initial states. This means that rest or motion at constant speed in a straight line are the normal conditions of particles, and if they are in one or the other of these states, then we know that no forces are acting on them. In either of these two possible states the world line of a particle in space-time would just be a straight line.

The mathematical definition of a straight line is that it is the shortest distance between two points. This is true if we are joining two points on a plane surface or in ordinary three-dimensional space, but it is not true if we are limited to moving on a curved surface or within a confined region of space that has curvature. For boats and ships traveling on the surface of the sea, which is curved, the curvature has to be taken into account when planning a route. An aircraft is confined to the atmosphere of Earth, the altitude of which is small compared to the radius of Earth, so once again we have to take into account the curvature of the atmosphere when flying over large distances. Pilots and the navigators of ships are very much aware of this. In going from one port to another distant port a few thousand miles away, seamen know that they have to sail along what are called *arcs of great circles*—a great circle being one that cuts the surface of Earth exactly in half. Such a circle is a special case of a class of mathematical curves called *geodesics*, and they represent the shortest distance between two points on a curved surface or in a curved space. We can use this concept to discuss one of the consequences of the general theory.

The general theory sees motion under gravity as being "normal" motion, so we do not have to look for other forces unless a body is moving differently from its normal motion. However, its world line will only be straight if is a long way from any massive object. According to this theory, the "shape" of space-time near massive objects is not flat, but curved, and there, under the conditions of a curved space-time, particles will follow special curved paths, which will be geodesics. The curvature of space-time is determined by the presence and distribution of matter, and because energy has mass, it is also determined by the distribution of energy. The rules that allow us to calculate the curvature of space-time from the distribution of mass and energy are called the *field equations* of the theory.

Once we have calculated the curvature of space-time, using these field equations, we can calculate the geodesics of this space-time, and these will tell us how particles will move if they are subjected to no other forces, such as, for example, electricity and magnetism. A ray of light will, because information travels along it at the speed of light, be a special type of geodesic, called a *null-geodesic*. Far away from all massive objects, this null-geodesic will be a straight line. This means that over the vast distances between stars, we can treat light as if it is traveling in straight lines, to a high degree of approximation. This is no longer true near massive objects.

Near the Earth, which is much less massive than the sun, the curvature of a light beam is very slight. However, the effect on a ray of light just grazing the surface of the sun can be detected, in appropriate circumstances. As we saw in the last chapter, such a situation arises during a total eclipse. Suppose that, at a given time of year, a particular star is just behind the sun, so that some of the light from this star will just graze its surface, and then reach the surface of the Earth. In effect, we will be able to see just behind the sun. Under normal circumstances the brightness of the sun would prevent us from seeing the star, but during a total eclipse, the moon will come between us and the sun, thus blotting out its rays, and so we will be able to see the star. Because the moon is much less massive than the sun, its own effect will be negligible. In Chapter 5 we discussed the first time such observations were made on the occasion of the total eclipse of May 29, 1919. These observations showed that Einstein's general theory of relativity was a much better theory than that of Newton.

The Origins of Quantum Theory

Quantum theory has its origins in the work of Max Planck (1858–1947), a German physicist who presented an important paper on the subject in 1900. Planck was trying to explain the shape of what is called the *black-body spectrum*. We are all familiar with such phrases as *red hot* or *white hot*, which emphasize the relationship between the color of the light emitted by an object and its temperature. Objects that are white-hot are much hotter than those that are red-hot. Most bodies broadcast energy in a wide range of wavelengths, from X-rays to very long radio waves. However, at a given temperature, most of the radiation given off will be close to a specific wavelength; that is, there is a specific relationship between the wavelength at which most of the energy is emitted, and the temperature. A mathematical formula representing this relationship was first discovered by the German physicist Wilhelm Wien (1864–1928), so it is called *Wien's law*. It tells us that the wavelength at which most of the radiation is emitted gets shorter as the temperature of the body increases, so, because red light is of a longer wavelength than blue light, an object that emits red light will be cooler, relatively speaking, than one that emits blue light. Cooler objects will emit most of their radiation in the infrared part of the spectrum, which our eyes cannot see. Cooler objects will still emit most of their radiation as long radio waves. A very important feature of these results is that they do not depend on the material of which the objects are made.

The general shape of the curve that relates energy emitted to wavelength is called the *black-body spectrum*. Before Planck did his research, physicists had been unable to understand the theoretical reasons for the shape of this curve. It was assumed that the energy carried by the radiation of a given wavelength could have any value; that is, that the amount of energy could vary in a continuous way. What Planck did was to propose that radiation energy had to be discrete—that the energy had to come in little packets, and the size of the smallest packet depended on the wavelength, the speed of light and a constant, called *Planck's constant*. Such a packet of energy is called a *quantum of radiation*, and collections of such packets are called *quanta*. Using the concept of quanta, Planck was able to work out the shape of the black-body curve, and so we now call this curve simply the *Planck Curve*.

Einstein and Quantum Theory

Another problem existed in physics at the end of the 19th century, involving a phenomenon called the *photoelectric effect*: When light strikes the surface of certain metals, in a vacuum, it was observed that electrons could be ejected from the surface. This could not be explained if the energy carried by light varied continuously. Einstein saw the importance of Planck's quanta for explaining this phenomenon. If one assumed that the little packets of light energy behaved as if they were particles, then a particle or quantum of light could collide with an electron near the surface of the metal, and, as a result, the electron would be emitted from the surface.

Quantum Theory and the Bohr Atom

The Danish physicist Niels Bohr (1885–1962) successfully applied quantum theory to exploring the structure of the atom and the specific set of spectral lines emitted by atoms of different elements. The model of the atom proposed by Rutherford consisted of a dense central nucleus, which had a positive electric charge, around which orbited a number of electrons. In a neutral atom, the number of electrons orbiting the nucleus will be equal to the number of positive charges in the nucleus. According to Maxwell's classical theory of electromagnetism, the electrons doing this orbiting would be radiating electromagnetic waves continuously, and because this meant they would be losing energy, they would spiral into the central nucleus. Bohr applied quantum theory to the problem. By doing so he was able to show that the electrons in the hydrogen atom could only orbit the nucleus in certain allowed orbits that were specific distances from the central nucleus. When an electron is orbiting in any one of these orbits, it does not radiate. However, it does give off radiation, of a definite wavelength, when it moves from an orbit far from the nucleus to one closer in. In other words, it does not radiate while it is in a stable orbit, but it radiates when moving from one orbit to another. It could also absorb radiation of a definite wavelength, when it moves from an orbit closer to the nucleus to one further out. It is this behavior of electrons that give rise to the observed set of spectral lines that could be used to identify the atoms of different chemical elements, because the charge on the central nucleus and the number of electrons orbiting it will be different for each element.

De Broglie Waves and the Bohr Atom

Prince Louis de Broglie (1892–1987) was a French nobleman who made an important contribution to early quantum theory. Just as Planck and Einstein had shown that light sometimes behaved as if it were made up of particles, so de Broglie proposed that particles could sometimes behave as if they were waves. In his paper, de Broglie proposed, not that the particles sometimes "turned into" waves, but that "pilot waves" guided the motion of particles, and, hence, under certain circumstances they would demonstrate the type of behavior that one normally associated with waves.

These waves were not at all like electromagnetic waves, and they were sometimes referred to as *matter waves.* De Broglie was also aware that these waves had to propagate through space faster than the speed of light. This was necessary because the pilot waves acted as if they were guiding the particle, informing it where it had to go. Thus, to get ahead of a particle traveling just slower than the speed of light, these waves had to travel faster than light, in order to probe the region of space ahead of it. An analogy may help to explain this: The navigator of a supersonic jet plane, flying at night, or in fog, has to use the radar of the aircraft (which uses radio waves traveling at the speed of light) to probe the space ahead of the plane. It is not possible to use sound waves for this purpose, as bats do, because the plane is traveling faster than sound.

De Broglie went on to point out that this did not conflict with the theory of relativity, because it was their "phase velocity" that was greater than that of light, whereas energy was transported at the "group" velocity, which was always less than the speed of light. The concepts of phase velocity and group velocity can easily be understood with reference to the expanding ring of waves that propagates outward when a stone is dropped into a pond: The ring itself will, on careful inspection, be seen to consist of a number of smaller wavelets, which start on the inner edge, move through the ring, and decay on the outer edge. These wavelets will be seen to be moving faster than the ring as a whole. They are moving with their respective phase velocities, whereas the ring as a whole moves with the group velocity.

Schrödinger and the Structure of the Atom

The work of de Broglie in some ways represented a transition in the approaches to quantum mechanics. It gave rise to the work of Austrian physicist Erwin Schrödinger (1887–1961), and this work in turn gave rise to a much more formal and mathematical approach to the application of quantum theory to physical problems. The contributions made by Planck, Einstein, and de Broglie were based more on physical intuition than on mathematical sophistication. Schrödinger's work was to change all that.

It has been said that Schrödinger first heard about the work of de Broglie in a reference in one of Einstein's papers. He decided it might be possible to improve on the Bohr model of the atom by taking the wave nature of the electron into account. The de Broglie wavelength associated with a moving electron depended, among other quantities, on its speed. Schrödinger was able to show that the Bohr orbits corresponded to speeds (which also depended on the size of the orbits) that gave rise to wavelengths that would fit a whole-number of times into the circumference of the orbits. Other orbits were not allowed because the associated wavelengths could not fit a whole-number of times into their circumferences.

All waves, from electromagnetic waves to water waves, are governed by a class of mathematical formulae called *wave equations*. Schrödinger generalized the work of de Broglie to derive such an equation, which is now called *Schrödinger's wave equation*.

Further Developments in Quantum Mechanics

The German physicist Arnold Sommerfeld (1868–1951), in 1916, generalized the work of Bohr, who restricted the electron orbits in the atom to circular ones to include elliptical orbits, and in doing so he had to include the results of Einstein's special theory of relativity. This was largely because the electron, in going around the central nucleus in elliptical orbits, was moving faster when close to the nucleus than when it was far from the nucleus, and this meant that the mass of the electron would change according to Einstein's formula. Although Schrödinger attempted to include special relativity in his wave equation, he was not successful in doing so.

It was left to the British physicist Paul Dirac (1902—1984) to succeed in including relativity in an equation that described the motion of subatomic particles. This is naturally known as the *Dirac equation*, and it showed not only that the spin of the electron (that the electron behaved as if it was spinning on its own axis) was a necessary consequence of the special theory of relativity, but also that there should exist particles with the same mass as the electron, but with a positive charge. A particle having these properties was discovered two years later by the American physicist Carl Anderson (1905–1991), and this discovery was later confirmed by Patrick Blackett (1897–1974), a British physicist.

As an undergraduate at Manchester University, my first introduction to the work of Sommerfeld and Dirac came from my lecturers and a textbook called *Fundamentals of Modern Physics* by Robert Martin Eisberg, who was at that time associate professor of Physics, University of California, Santa Barbara. I have always remembered a statement he made in that book:

> Therefore the results of a complete relativistic treatment by the Dirac theory can be expresses by the single equation.... [I have omitted the mathematical details.] These results are in exact agreement with the predictions of Sommerfeld's theory. Since the Sommerfeld theory was based on the Bohr theory, it is a very rough approximation to physical reality. In contrast, the Dirac theory represents an extremely refined expression to our understanding of physical reality. From this point of view the agreement between [the two equations] is one of the most amazing coincidences to be found in the study of physics.

At the time, I agreed with Eisberg, but through the years I have began to think otherwise, particularly when I saw a mathematical derivation of Schrödinger's equation by Richard Feynman, using a very different approach from the one used by Schrödinger. I now believe that, in many instances, more pictorial approaches can be just as good as mathematical models in arriving at verifiable formulae that can be used from computations.

Even before Schrödinger had done his work on quantum theory, another approach had been worked out by Werner Heisenberg (1901–1976). This highly mathematical approach was called matrix mechanics. In mathematics, a matrix is a set of quatities in a rectangular array, which can be used in certain mathematical operations. The array is usually enclosed in square brackets or large parentheses. Heisenberg used these mathematical structures to discuss his own formulation of quantum mechanics. His work was further developed by himself, Max Born (1882–1970), and Pascual Jordan (1902–1980). In 1926, Schrödinger showed that matrix mechanics was equivalent to his own wave mechanics. Heisenberg is best known for the uncertainty principle that bears his name. According to this principle, it is impossible to know both the position and the momentum (which is the speed multiplied by mass) of a subatomic particle at the same time. If one makes an accurate measurement of the speed of a particle, then the act of measurement will make the measurement of its position uncertain.

The concept can be understood by means of a simple analogy: A police radar trap works by bouncing radio waves off cars, and then measuring the change in wavelength that arises from the motion of the car, using the Doppler effect. The momentum in the radio beam is small in comparison with the momentum of the car, so the bouncing of the beam off the car will only change its speed by a small amount. However, in order to measure the speed of the electron, with sufficient accuracy, one will have to use very short radio waves, and the energy in the beam will deflect the electron, so we will know very little about its subsequent position.

Many physicists, including Einstein, did not like the uncertainty of quantum mechanics, nor did they like the statistical interpretation that some physicists claimed should be given to the wave functions associated with quantum mechanics. Einstein is reputed to have said that he did not believe that "God played dice with the universe." It was their dissatisfaction with these aspects of quantum theory that led physicists to propose that the theory could be made deterministic by the inclusion of "hidden variables." I will discuss this aspect of the subject, very briefly, in the next section.

Hidden Variables in Quantum Theory

Hidden variables has been defined as any set of hypothetical physical quantities, knowledge of whose values would permit more precise predictions of the results of measurements on a system than the statistical predictions of quantum theory. The normal statistical interpretation of quantum theory rules out the possibility of such hidden variables. Hungarian-American mathematician John von Neumann (1903–1957) produced a mathematical proof that was supposed to show conclusively that hidden variables were inconsistent with quantum theory. His standing in mathematical physics was so high that this proof was taken as the final word on the subject by most physicists—but not all. In 1952, David Bohm showed that it was possible to understand the results of Schrödinger's non-relativistic quantum mechanics by introducing a "hidden variable" in the form of a "quantum potential." The quantum potential had one surprising feature: similar to de Broglie's pilot waves, it had to propagate faster than light because it carried information on the region ahead of the particle. In 1966 John Bell (1928–1990), a theoretical physicist from CERN in Geneva, proved a theorem, now called *Bell's theorem*, which shows that (as he said in *Speakable and Unspeakable in Quantum Mechanics*), in quantum theory, "an explicit causal mechanism exits whereby the disposition of one piece of apparatus affects the results obtained with a distant piece."

Bell's theorem demonstrates that if two particles have interacted in the past, then each particle carries a memory of that interaction, which can be instantaneously recalled, so subsequent measurements on the pair will always be correlated. To quote from *The Cosmic Blueprint* by Paul Davies:

> Bell showed that quantum mechanics predicts a significantly greater degree of correlation than can possibly be accounted for by any theory that treats the particles as independently real and subject to locality. It is as if the two particles engage in a conspiracy to cooperate when measurements are performed on them independently, even when these measurements are made simultaneously. The theory of relativity, however, forbids any sort of instant signaling or interaction to pass between the two particles. There seems to be mystery, therefore, about how the conspiracy is established.

Chapter Seven

Theories on the Nature of Matter

In this chapter I will discuss theories concerned with the nature of matter. I will start with a brief overview of the quark theory of matter, and then we will discuss string theory, concentrating on the problems posed by this approach, which at one time was seen as "a theory of everything." The major part of this chapter is a presentation of my own plasma space theory of matter. The chapter will end with an explanation of how my own theory can deal with the problems posed by dark energy and dark matter.

The Quark Theory of Matter

When subatomic particles were collided with each other, at enormous energies, in large machines called *particle accelerators*, many new particles began to appear. Most of these particles are short-lived, and they have different masses, electric charges, and spins. Charge and spin seemed to come only in whole numbers of certain quantities or halves of these quantities, so they also came to be known as *quantum numbers*. It soon emerged that more quantities were required to classify these particles on the basis of their behavior. Most of these other quantities cannot be easily associated with physical concepts, such as charge or spin, so they were given abstract names such as *hypercharge* and *isospin*, but they also generally came in integers and half-integers of certain quantities, so they are also called *quantum numbers*.

Using these numbers, and arranging the particles according to the values of their quantum numbers, it soon became apparent that they fell into very neat patterns. Scientists were surprised by these patterns, but they were even more surprised to find that they can be represented in a very simple way, if one assumes that all members of a particular class of particle—called *hadrons*—were made up of a small number of elementary entities. These entities, now called *quarks*, were postulated by M. Gell-Mann and G. Zweig in 1963. Independently, these two physicists succeeded in accounting for all the regularities by assigning appropriate quantum numbers to three quarks and their antiquarks. Putting these quarks together in various combinations, they were also able to form other classes of particles—called *baryons* and *mesons*—whose quantum numbers were simply obtained by adding those of the constituent quarks. Baryons, in this scheme, consisted of three quarks, and their corresponding antiparticles of the corresponding antiquarks, and mesons of a quark plus an antiquark.

Although the quark model continues to be very successful in accounting for the regularities in the particle world, it is no longer used in its original simple form, and particle physicists have had to introduce additional quarks to account for the large variety of hadron patterns. The simple quark model leads to severe difficulties, in spite of its efficiency and simplicity, if the quarks are considered to be the physical constituents of hadrons. The forces that hold these particles together are supposed to be carried by other particles exchanged between the interacting particles. This means that these exchange particles must also be present inside hadrons. If this were indeed the case, then they too would contribute to the properties of the hadrons, and this would destroy the simple additive scheme of the quark model. These are some of the problems that led some physicists to propose an alternative approach to high-energy physics.

String Theory

The most recent review of the history of string theory is the book by Lee Smolin called *The Trouble with Physics: The Rise of String Theory, the Fall of Science, and What Comes Next.* It is appropriately subtitled, because the book highlights the major problems with the theory. In the introduction, Smolin gives a brief review of some recent views on the subject, and a short summary of the theory: "It purports to correctly describe the big and the small—both gravity and the elementary particles—and to do so, it makes the boldest hypotheses of all the theories: It posits that the world contains as yet unseen dimensions and many more particles than are presently known.... It claims to be the one theory that unifies **all** the particles and **all** the forces in nature."

In the February 2007 issue of *Physics World,* Smolin's book was reviewed by Michael Riordan, who teaches the history of physics and technology at Stamford University and the University of California, Santa Cruz, who said this: "But string theory is not really a 'theory' at all—at least not in the strict sense that scientists generally use the term. It is instead a dense, weedy thicket of hypotheses and conjectures badly in need of pruning."

When I wrote my book *The Paranormal: Beyond Sensory Science* in 1992 (reissued in 2003 as *The Third Level of Reality*), I quoted two opposing views about string theory, both from Nobel Prize–winning physicists (from Paul Davies's *Superstrings*). The first was from Richard Feynman: "It is precise mathematically, but the mathematics is far too difficult for the individuals who are doing it, and they don't draw their conclusions with any rigor. So they just guess." The other is from Steven Weinberg: "This is physics in a realm that is not directly accessible to experiment, and the guiding principle can't be physical intuition because we don't have any intuition for dealing with this scale. The theory has to be conditioned by mathematical consistency."

We see that the same general arguments between the opposing camps are still raging after 15 years!

The Plasma Space Theory of Matter

An ordinary gas consists of neutral atoms in which the number of protons in the nucleus is exactly balanced by the number of electrons orbiting the nucleus. The movement of such atoms is not influenced by magnetic fields. A plasma is a very hot gas, in which the atoms are moving so fast that some of their electrons have been stripped away by collisions with other atoms. Such atoms are electrically charged, and they are called *ions*. These ions and electrons do interact with magnetic fields, in that they will spiral around the lines of force used to represent the direction of the magnetic field. They will thus behave as if they are "threaded" to the lines of force in a way not dissimilar from the way beads are threaded to the string of a necklace. If the ions and the electrons are moved, then the lines of force will also move, and if the lines of forced are moved, so will the ions and the electrons. I am proposing that there are two different types of space: ordinary space and plasma space. The points of ordinary space are rather like neutral atoms in that they are not threaded by lines of force. The points in plasma space *are* threaded by lines of force.

The plasma space theory of matter proposes an analogy between the ions of a material plasma and the points of plasma space, and between the magnetic lines of force that thread the plasma and the electric lines of force that thread their way through plasma space. It is further proposed that, just as there are no ends to magnetic lines of force, there are no ends to electric lines of force. The electric lines of force are completely frozen into plasma space from the beginning of time. The basic particle of matter in this theory is the neutron. The core of the neutron is a force-free ball of yarn configuration of electric lines of force, and the electrical energy compacted within this ball of yarn gives the neutron its mass. When the neutron decays, it is the outer unstable cloak of lines of force that rearranges itself to give the proton-electron configuration and a neutrino. There are then three stable particles: proton, electron, and neutrino. The core of each of these is a force-free ball of yarn configuration of electric lines of force, embedded in the plasma space confined within these particles. The neutrino has no charge, so it is just a compacted ball of yarn of electrified plasma space.

The decay of the neutron resembles, to some extent, the formation of a solar prominence and a sunspot pair. When it decays, the outer shell of the neutron, because it is unstable once it is outside the nucleus, will rearrange itself in such a way that the lines will splay outward from the neutron to form a proton, wrap themselves around the core of the new electron, which has been ejected, and will then twist back to the proton in an extremely thin bundle of electrified plasma space. This thin bundle will be encased in a sheath of insulating space, which has elastic properties. One end of this sheath will be anchored on the electron, and the other on the proton. The insulating properties of the sheath will prevent the lines of force from leaking out. In the process of forming the electron and the proton, a very small ball of yarn configuration (which is also force-free) of electrified plasma space is produced, of extremely small size, and hence very little mass—this is the neutrino (see Figure 7.1). The sheath of insulating space that takes the electric lines of force back from the electron to the proton, and the lines of force on the inside of the sheath, are able to stretch to cosmological lengths. Initially the lines of force on the inside resemble a force-free cylindrical configuration (see Figure 7.2), but the farther the electron is separated from the proton, the more they become nearly parallel to each other. However, the elastic properties of the sheath prevent then from bulging out due to the electrical pressure of parallel lines going in the same direction. To start with, the elasticity of the sheath, along its length, is very weak, but the tension in it increases as the distance between the electron and its parent proton increases.

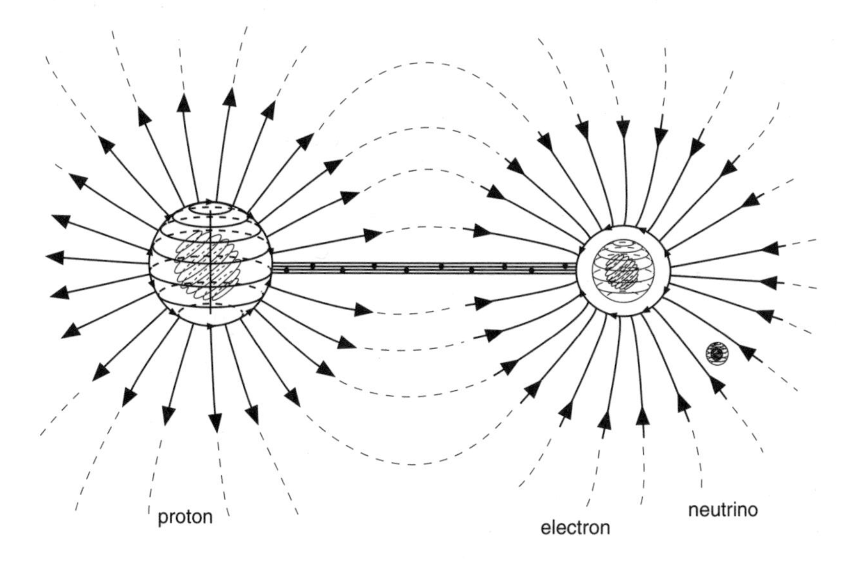

Figure 7.1
The plasma space configuration of a proton, electron, and neutron.

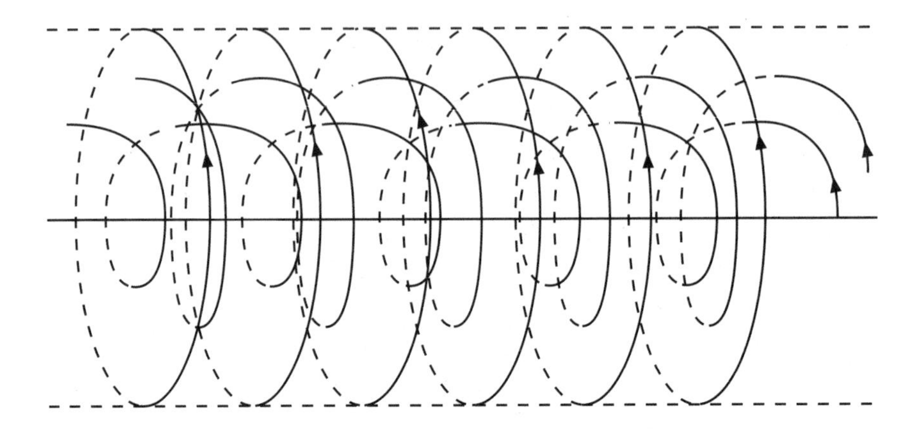

Figure 7.2
The force-free plasma space configuration linking the electron back to the proton.

The ball of yarn configurations within the electron, proton, and neutron (either for the short time it is free or within the nucleus) are all spinning on their own axes, and thus they will behave as little magnets. It is this behavior that gives rise to the measurable "magnetic moments" of these particles. The difference between these particles and their corresponding antiparticles lies in the difference between the spinning of these yarn balls and the direction in which the lines of electric force are wound on the outer layers of each yarn. For antiparticles it is in the opposite direction of what it is for particles.

Plasma Space Theory and the Evolution of the Universe

Before the big bang, all the electrified plasma space was confined to a very small volume. It was the electrical pressure in this volume that led to the expansion of the universe. Initially, after some expansion, the electrical plasma space fragmented into a large number of neutrons, which could exist because they were still in close proximity to each other. It is just as if, in the global village of the early universe, a short-term stable marriage could exist between an electron and a proton, and the neutrino was the wedding ring of the marriage. Originally there was just a tangled loop of electrified plasma space, so a large number of reconnections had to take place to produce the neutrons. This process, together with the unstable vibrations of the original tangled loops, produced the high temperature of the early universe. With the further expansion and subsequent dispersion of the global village, the neutron marriages dissolved, and the neutrinos produced as a result were the discarded wedding rings. At a much later stage in the universe, neutron marriages could again exist in the much smaller villages of stable nuclei.

The eventual recycling of matter in the early universe and the later formation of galaxies, stars, and planets has meant that the electrons have become separated from the protons to which they were once married. A graphic description of the type of recycling that takes place in stars is to be found in Sir Arthur Eddington's *The Internal Constitution of Stars*:

> The inside of a star is a hurly-burly of atoms, electrons and ether waves. We have to call to aid the most recent discoveries of atomic physics to follow the intricacies of the dance.... Try to picture the tumult! Disheveled atoms tear along at fifty miles a second with only a few tatters left of the elaborate cloaks of electrons torn from them in the scrimmage. The lost electrons are speeding a hundred times faster to find new resting-places. Look out! There is nearly a collision as an electron approaches an atomic nucleus; but putting on speed it sweeps round it in a sharp curve. A thousand narrow shaves happen to the electron in 10^{-10} of a second; sometimes there is a side-slip at the curve, but the electron still goes on with increased or decreased energy. Then comes a worse slip than usual; the electron is fairly caught and attached to the atom, and its career of freedom is at an end. But only for an instant. Barely has the atom rearranged the new scalp on its girdle when a quantum of ether waves runs into it. With a great explosion the electron is off again for further adventures. Elsewhere two of the atoms are meeting full tilt and rebounding, with further disaster to their scanty remains of vesture.

Eddington's description is a classic piece of science writing. Even in his day the concept of the ether had already been replaced by that of the electromagnetic radiation field. He was using the concept as a literary device.

Thus where he used the term *ether waves* it should be understood that he is referring to electromagnetic waves.

The electric lines of force go from the proton to the electron in normal space, but they go from the electron back to the proton in the opposite direction in plasma space, which it encased in insulating sheaths. This means that as an electron or proton describes its world line in space-time, this world line has a physical reality in the form of a fiber bundled of electrified space surrounded by a sheath of elasticated insulating space. The recycling and evolution of matter means that the whole of space is criss-crossed by an interconnecting and interwoven web of such world lines, rather like a disordered spider's web.

The Laws of Normal Space and Plasma Space

The normal space that we investigate with scientific equipment, and is known to us through our senses, is the space between the threads of the web. The measured speed of light is the limiting speed in normal space. Interactions between particles whose world lines have crossed in the past can take place faster than the speed of light; the particles retain memories of their past interactions. However, it is only recent memories that are retained unambiguously. The older memories are degraded by the number of interactions that lie between them and the present. The separate forms of matter are nothing more than colonies within the web. As the universe expands, so the lines of electric force of the web and the elasticated insulating sheaths that restrain them are stretched, and the sizes of the spaces between them will also get bigger. The large-scale colonies of matter have mass as a result of the localized energy embedded in them, and as they move through the web of world-lines they merely brush them aside, or trail them behind, without interacting with them in any other way. Large masses thus do not respond to the continuous and eternal vibrations of the world-line web, and so they obey the laws of Newtonian physics at low speeds and the laws of special relativity at speeds approaching that of light. Subatomic particles do feel the vibrations of the web, and as a result they obey a rather different set of laws. These are the laws of quantum mechanics.

Professor David Bohm, in *Wholeness and the Implicate Order*, attributed quantum effects to the fluctuations of a special field, which he called the *psi-field*: "The fluctuations of the psi-field can be regarded as coming from a deeper sub-quantum level, in much the same way that fluctuations in the Brownian motion of a microscopic liquid droplet come from a deeper atomic level."

Brownian motion is the incessant movement of tiny particles of dirt or dust suspended in a liquid. The movement of these dust particles, which is random in direction, can be seen in a droplet of liquid when viewed under a microscope. It arises from the bombarding of particles of dust, which are larger than atoms, by the atoms themselves. The phenomenon was first discovered by Robert Brown in 1827, and it was treated theoretically by Albert Einstein in the first decade of the 20th century.

The plasma space theory of matter says that the fluctuations of the psi-field are due to the world line web.

The Polynesian Navigators of the Quantum World

The last few pages may have been difficult to follow. Here is an analogy that should help to clarify some consequences of the theory I am proposing: The seamen of Polynesia navigate their way around the Polynesian island using a variety of navigational aids, one of which makes use of the pattern of ocean swells that exist around the islands. The waves coming from the open ocean will be reflected from the beaches of the nearby mainland, but when there is an island off the mainland, the incoming waves and the reflected waves will both be distorted by the presence of the island, and the interaction between the two sets will give rise to an interference pattern around the island. If there are a number of islands, then there will be an even more complex interference pattern around them. The young navigators of the islands are taught how to use the bobbing up and down of their boat, together with a map of the interference pattern, to find their approximate position with respect to the islands.

In an ordinary plasma, consisting of neutral atoms (or molecules) and ions, there are two types of waves that propagate through it: ordinary sound waves, which propagate in all directions, and which involve atoms and ions, and Alfven waves. As we already saw in Chapter 1, Michael Faraday believed that magnetic lines behave as if they have a tension along their lengths (the concept was given mathematical substance by the work of Hannes Alfven). This means that if one plucks a field line, it will vibrate, rather like the plucked string of a violin. It is this vibration that we call an *Alfven wave*, and it propagates along the field line with the Alfven speed. The speed of sound is related to the pressure and density of the plasma, but the Alfven speed is related to the strength of the magnetic field and the density of the plasma. The two speeds are, more often than not, different. In a strong magnetic field, the Alfven speed can be much higher than the speed of sound. It is also confined to where there are field lines. I am proposing that ordinary electromagnetic waves propagate through ordinary space at the speed of light, but Alfven plasma space waves can propagate along the electrified world lines of the world-line web that permeates the cosmos. The speed of Alfven plasma space waves can exceed the speed of light, and thus these are the pilot waves, first proposed by Prince Louis de Broglie, which inform subatomic particles where to go.

Plasma Space and the Elementary Particles

Geophysicists working on the theory of how the Earth's magnetic field may be generated by the motions of the fluid parts of its interior, describe the structure of the field, and the flow lines of the fluid's motion, in terms of a special type of mathematical function called *spherical harmonics*. This type of function can also be used to describe the shape of a sphere that has been distorted in any way. Simple shapes can be represented by spherical harmonics that have just one or two maxima or minima associated with them. These are called *low-order harmonics*. More complex shapes use harmonics with several maxima and minima—these are called *high-order harmonics* (see Figure 7.3). These harmonics were first introduced into mathematics by the French mathematician Pierre Laplace (1749–1827), who built on the work of another French mathematician, Legendre. Both men were trying to find a way of working out the gravitational field of planets that were not exactly spherical. These functions are now used in many different areas of applied mathematics to describe the shapes of the planets and the structure of their gravitational and magnetic fields. On a much larger scale, while doing research for my master of science and for my doctor of philosophy degrees, I pioneered the use of spherical harmonics to discuss the structure and symmetry properties of the vast magnetic field that threads its way

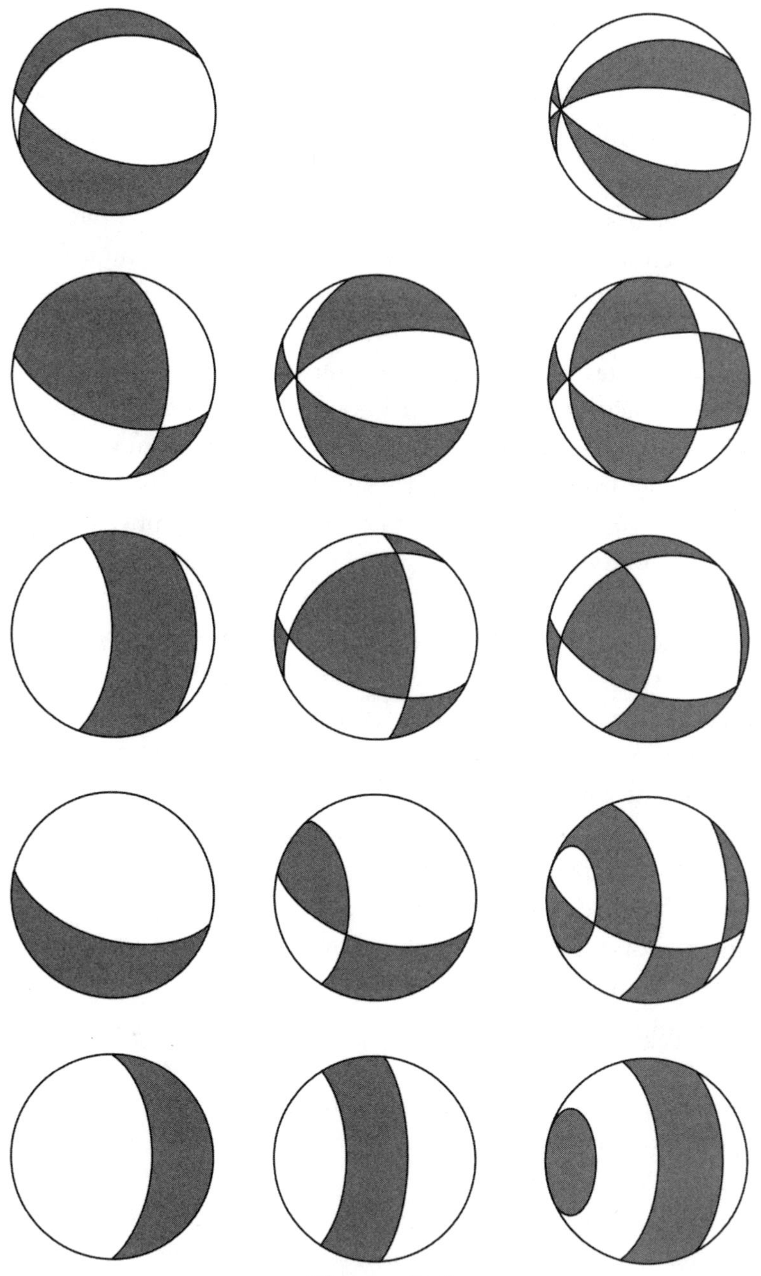

Figure 7.3

A pictorial representation of some of the spherical harmonies.
The dark regions on the spheres are concave,
and the paler regions are convex.

through the whole Milky Way galaxy. On a much smaller scale they are used to describe the structure of the atom in terms of quantum-mechanical wave functions. The use of these harmonics in atomic structure has yielded many results with far-reaching consequences. One topic of particular interest is that of forbidden lines.

When the older type of quantum mechanics was first applied to atomic structure, each state of the atom was described by a small number of quantum numbers. It was soon discovered that atoms were generally not allowed to go from certain states to certain other states, and the spectral lines that would have resulted from such transitions were never observed in the laboratory. These transitions became known as *forbidden transitions*, and this concept led to the idea of "selection rules" that quantum numbers had to obey in going from one state to another. With the development of Schrödinger's model of the atom, it became clear that spherical harmonics were necessary to describe the wave functions associated with the model, and further investigations showed that the forbidden transitions and selection rules followed directly from the mathematical properties of the spherical harmonics. It also clarified the real meaning of a forbidden transition. It showed that they were linked to states that had extremely long "lifetimes," and that, in the high densities of terrestrial laboratories, atoms would undergo collisions with other atoms before they were allowed to decay from these states, and as a result they would be excited into a different state, with a shorter lifetime, from which they could decay more rapidly. However, in the extremely low densities and large volumes that are quite often encountered in gas clouds between the stars, it was possible to observe these forbidden transitions.

In extending the theory of plasma space to discussing elementary particles, the internal structure of the electric fields within protons, neutrons, and neutrinos can also be described in terms of spherical harmonics. Because protons and neutrons can both exist in the nuclei of atoms, they are given the collective name of *nucleons*. When nucleons were bombarded with electrons it seemed as if they had some internal structure, and this structure was attributed to the quarks we have already discussed. According to the plasma space theory of the elementary particles, the collision between an electron and a nucleon will excite the nucleon into states that can be described by higher-order harmonics, in a way that will give the impression that the nucleon has internal structure. Many collisions between a large variety of particles have revealed a plethora of so-called elementary particles, most of which are extremely short-lived. These particles can be described by sets of quantum numbers, and these numbers follow particular patterns, enabling high-energy physicists to group the particles into families. The plasma space theory of matter says that these short-lived particles are not particles at all—they are the result of collisions that excite known stable particles into states described by higher-order harmonics, which we will call *multipole extensions* of the original fields of these particles, and the subsequent fragmentation of such a multipole will give rise to short-lived particles that are not really elementary.

The work of Bullard and Gellman, on how internal motions of plasmas can convect magnetic fields to generate new magnetic configurations, showed that these fields also obey selection rules, and these, once again, arise from the mathematical properties of the spherical harmonics used to describe the field and the internal motions of the plasmas. They also showed that these selection rules gave rise to families of magnetic configurations that exhibited remarkable symmetry properties. The plasma space theory of elementary particles proposes that the quantum numbers associated with the discovered families of elementary particles, and the patterns that arise from these families, are the result of the mathematical properties of the spherical harmonics used to describe the internal fields of the colliding particles and their relative motions with respect to each other. The rules that govern particle collisions, and the symmetry properties of the resulting families of particles, can once again be considered to arise from the mathematical properties of the spherical harmonics. The ancient concept of "the music of the spheres" seems to manifest itself at a variety of levels in the physical universe, from the very large to the very small.

Nuclei of the Heavier Chemical Elements

The nucleus of ordinary hydrogen just consists of one proton. However, hydrogen comes in three different species: ordinary hydrogen, deuterium, and tritium. These are called the *isotopes* of hydrogen. The chemical properties of these isotopes are determined by the one electron orbiting the central nucleus, and, as all three isotopes have the same electronic structure, they have the same chemical properties. However, the atom of deuterium has about twice the mass of ordinary hydrogen, because it has a neutron in its nucleus, in addition to the proton. The mass of the tritium atom is about three times that of ordinary hydrogen, because in its nucleus it has one proton and two neutrons.

The next chemical element in the periodic table is helium. Its nucleus has two protons and two neutrons, and, in the case of a neutral atom of helium, it has two electrons orbiting the nucleus. According to my theory, the positive electric charge of the nucleus, which is twice that of the charge on a single proton, comes from the rearranging of the outer cloak of lines of force from each proton to make a combined shield around the force-free ball of yarn configurations of the neutrons and the cores of the protons. The so-called strong nuclear force is due to the way these internal balls of yarn rearrange themselves to form a stable nucleus (Figure 7.4). The same type of reasoning holds for all the atoms of the other heavier chemical elements. For certain isotopes of these other elements the nucleus is not stable, which means some rearrangement has to occur, and in the process an electron, a positron, or an alpha particle (which is, in essence, the nucleus of the helium atom) will be ejected from the nucleus. Such isotopes are called *radioactive isotopes.*

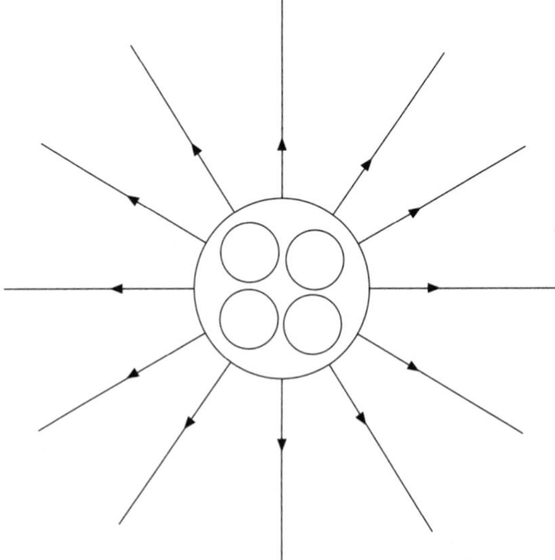

Figure 7.4
The lines of force radiating from a complex nucleus and the force-free windings within the nucleus.

BELL'S THEOREM AND QUANTUM MECHANICS

Bell's theorem basically shows us that if quantum mechanics is valid (and all physical experiments have so far failed to reveal that it isn't), then measurements made on two particles will always be correlated, no matter how far apart they are. This can be illustrated by using two subatomic particles. We have already seen that such particles spin on their own axes, rather as tops or planets do. Physicists call this property of particles their spin. Suppose we have a two-particle system in which one particle is spinning in the opposite direction of the other, when they are very close together—this is normally referred to as the one particle having spin up and the other particle spin down. If we measure the spins of the two particles after they have been separated by a large distance, then we find that the spins of the particles are still one up and one down.

Such particles, because of their spin, behave as if they were little magnets, so we say they have magnetic moments. It is possible to change their orientation by passing them through a magnetic field. Quantum mechanics tells us that if we change the orientation of one particle, so that instead of spinning up about a vertical axis, it is now spinning left about a horizontal axis, then we will find that the other particle is also spinning about a horizontal axis, but in the opposite direction, which we will call *right*. These results of quantum mechanics have been confirmed by two experiments, the first one performed in 1972 by John Clauser and Stuart Freeman in America, and the second by A. Aspect, P. Grangier, and C. Roger at CERN in Geneva, in 1981. Thus, remarkable though it may seem, there is some form of very rapid communication (faster than the speed of light) between the two particles.

TWO-PARTICLE INTERACTIONS

We have already seen that the balls of yarn of plasma space that form the cores of the proton and the electron are spinning on their own axes, and hence these particles will behave as little magnets with their own spin and magnetic moments. Spin is also called *angular momentum*. To find the angular momentum of a single particle moving in a circle about the center of a circle, we have to multiply its ordinary momentum (which is its speed multiplied by its mass) by the distance the particle is from the center. Angular momentum is always conserved in physical systems. Thus, if we change the distance of the particle from the center about which it is moving, then the speed with which it is moving must also change: if the distance were increased, the speed would decrease, and vice-versa. This point is well-made by an ice dancer: If she starts spinning with her arms outstretched and then lowers her arms, she will begin to spin faster. With the arms outstretched there is a certain angular momentum in the dancer, which will be conserved, and therefore as her arms are brought closer to the spin axis, her body will inevitably spin faster. This would be even more pronounced if she had small dumbbells in her hands.

If two subatomic particles have the same mass, and originate at the same point in space at the same time, but one is spin up and the other spin down, then, as we have already seen, quantum mechanics tells us that they will retain this correlation no matter how far apart they are. In other words, if some of the fiber bundles of the web are twisted at one point, then this twisting will be communicated at very high speed (faster than light) to another part of the web, provided there is a clear channel between the two parts—that they share a branch point in space-time. For large bodies and collections of particles this will, in general, not be the case, because they are composed of particles that have more complex histories, and hence the channel will be noisy, and the communication will not be very clear. It is for this reason that, according to plasma space theory, it would not be possible to develop a system of communication using the very rapid transfer of angular momentum between parts of the world-line web.

The Wave-Like Nature of Electrons

We have already looked at the interference that can take place between two different sets of waves; we saw that Polynesian navigators can use interference between incoming ocean waves, the waves reflected by the coast, and the distortion of these two sets by islands to find their way around the islands. The interference of two sets of waves sent out from radio transmitters was also used by the navigators of ships (before the advent of satellite navigation) to find their way around certain parts of the world. Interference was used by the physicist Thomas Young to establish the wave nature of light. He passed monochromatic light (light having only one wavelength) from a source through a single slit to produce a coherent set of waves, and then passed the light through two further slits before allowing the combination to fall on a screen. The light waves from the two slits interfered with each other to produce a set of dark and light bands on the screen. A light band would be produced when the peaks and troughs of the waves, coming from the two slits, were in phase, and strengthened each other, and a dark band would be produced when the peaks and troughs were out of phase, and canceled each other out. This experiment was a convincing demonstration that light consisted of waves.

An experiment similar to Young's twin-slit experiment has been used to establish the wave nature of electrons. In a vacuum, the electrons from a filament were passed through two slits before being allowed to fall on a screen. The pattern produced on the screen resembled that produced by Young's slits in the case of light, thus showing that, under certain conditions, electrons behaved as if they were waves. When there was only one slit, then the electrons behaved as if they were particles, and formed a rather fuzzy line on the screen about the direction that passed from the filament through the single slit. In other words, most of the electrons passed through the slit in straight lines, as particles would, but a few did spread out after passing through, as one would expect if they retained some of their wavelike properties. The result of this experiment can be understood in terms of the world-line web.

The cosmic evolution and recycling of matter means, as we have already stated, that the whole of space is criss-crossed by the web. As the electrons move through space, they will undergo frequent collisions with the web, thus setting it vibrating. The electrons will also pick up some of the natural vibrations of the web. This is rather like a speedboat moving through the sea, which not only creates its own waves, but also responds to the natural waves of the ocean. The de Broglie wavelength of an electron depends on its momentum, and it is this wavelength that determines the interference pattern it will produce in a twin-slit experiment. The interaction of the waves produced by the motion of the electron and the natural oscillations of the psi-field give rise to a group of waves, the group velocity of which is equal to the velocity of the electron. When both slits are open, then the electrons passing through the two slits will give rise to waves in the world-line web that will interfere with each other, and the interference informs the electron where to go on the screen. The phase velocity of waves along the web can be greater than the speed of light, as required by Bohm's quantum potential and de Broglie's matter waves.

DARK MATTER AND DARK ENERGY

These two components make up the major part of the universe. According to the latest calculations, dark energy makes up 75 percent of the universe, dark matter 23 percent, and ordinary matter makes up only 2 percent of the universe. According to my theory of matter, in the early universe, the decay of neutrons to form protons, electrons, and neutrinos gave rise to myriads of thin bundles of electric lines of force encased in insulating space, and because of the turbulence then present, these bundles became very entangled. The elastic properties of insulating space mean that the tension in the sheaths that encase the electric bundles become more and more stretched. The dark matter comes from the energy within these stretched sheaths and the electric energy within the bundles. The stretched sheaths and the stretched electric bundles of force give an added source of attraction, which is partially the cause of the indications that there is more matter in galaxies and clusters of galaxies. The other source is the additional mass that results from the energy stored within the lines of force and in the stretched sheaths. Unlike the other invisible forces of nature, the tension in the elasticated sheath gets stronger as the distance between the galaxies increases. Thus at galactic and galactic-cluster dimensions, the tension acts like an additional attractive force. However, the situation changes when the distances of separation approach cosmological dimensions; then it becomes repulsive. This point was well made by Lee Smolin in *The Trouble With Physics*: "Most kinds of matter are under pressure, but the dark energy is under tension—that is, it pulls things together rather than pushes them apart.... But it turns out that when the negative pressure is negative enough, in general relativity it has the opposite effect. It causes the expansion of the universe to accelerate." Thus the behavior of the elasticated sheath contributes to dark matter when the tension is relatively small, but it becomes the main source for dark energy when the tension becomes very large.

The Five Great Problems

In his book *The Trouble with Physics*, Lee Smolin also discusses the five great problems in theoretical physics. His list is given below.

- Problem 1: Combine general relativity and quantum theory into a single theory that can claim to be the complete theory of nature.
- Problem 2: Resolve the problems in the foundations of quantum mechanics, either by making sense of the theory as it stands, or by inventing a new theory that does make sense.
- Problem 3: Determine whether or not the various particles and forces can be unified in a theory that explains them all as manifestations of a single fundamental entity.
- Problem 4: Explain how the values of the free constants in the standard model of particle physics are chosen by nature.
- Problem 5: Explain dark matter and dark energy.

Except for problem 4, my plasma space theory of the universe offers an explanation for these problems.

Just as the elegant mathematics of Maxwell's electromagnetic theory was based on the scaffolding provided by the pictorial and visual thinking of Faraday, and the full quantum theory was based on the scaffolding of the pictorial and visual thinking of Bohr and Sommerfeld, so I hope I am providing the scaffolding on which it might be possible to erect a fuller mathematical theory. (The rudiments of this theory are contained in a paper I wrote in 1989: "The Elementary Particles as Stable and Unstable Modes of Electrified Space-Time.")

Chapter Eight

Modeling Reality

In this final chapter we will look at different ways of modeling situations in the real world. Following a brief classification of the standard types of models and a short discussion of how they differ, we will consider two ways of modeling the reality of universe. The first we will call *galactic reality*, because it is based on our current understanding of our Milky Way galaxy, and the second we will call *urban reality*, because it is based on the operation of towns and cities.

Different Types of Models

There are four main types of models used by scientists and engineers: iconic, symbolic, mathematical, and analogous. Let us look at each type in turn.

Iconic Models

Iconic naturally comes from the word *icon*, which means "image" or "likeness." Thus an iconic model is similar to the real thing in every respect, although it may be a scaled-down version of the actual system being studied. Engineering prototypes are iconic models, and so are model steam traction engines and locomotives that work just like the real thing, but it is impossible to construct such models for natural systems. *Orreries* are mechanical models of the solar system, which simulate the movements of the planets around the sun and the moon around the Earth, but although they represent some of the essential features of the solar system, they fall down in many respects: They do not simulate elliptical orbits, all the planets move in the same plane, and mechanical arms, levers, and gears keep the planets in their orbits, rather than natural forces.

Symbolic Models

Symbolic models are among the most widely used class of models. Such models make no attempt at reality in a simple one-to-one way, but instead they make use of symbols to represent the actual objects of the system it is attempting to simulate. The circuit diagrams used by electrical, electronic, and radio engineers are examples of symbolic models. Large cities that have underground systems normally have symbolic maps or models that represent the essential features of the system, and can be used effectively by those who use the underground.

Mathematical Models

Mathematical models are precise symbolic models. When such models are applied to physical or astronomical systems, the symbols usually represent some measurable physical quantities. The relationship that exists between various physical quantities is represented by mathematical formulae. A mathematical model of the solar system must be able to predict the positions of the planets at any time in the future, from knowledge of their positions in the past.

Analogous Models

Analogous models are based on the analogies that exit between many apparently different systems in the physical universe. All working scientists make extensive use of such models. The outstanding American mathematician Professor G. Polya wrote several books on mathematical thinking. In *How to Solve It* he writes, "Analogy pervades all our thinking, our everyday speech and our trivial conclusions as well as artistic ways of expression and the highest scientific achievements. Analogy is used on very different levels. People often use vague, ambiguous, incomplete, or incompletely clarified analogies, but analogy may reach the level of mathematical precision. All sorts of analogy may play a role in the discovery of the solution, and we should not neglect any sort."

An important point about all models is that they are constructs of the human mind, and thus they have their own limitations. Einstein once said, in *The Meaning of Relativity*, "The only justification of our concepts and our system of concepts is that they serve to represent the complex of our experiences; beyond this they have no legitimacy."

ON THE PHILOSOPHY OF MODELING

Throughout the history of our attempts to model the physical universe there have been debates about those models that are supposed to represent reality and those that are mathematical constructions used to carry out calculations that can lead to testable predictions. Thus, in ancient Greece there was a distinction between Aristotle's model of spheres nestling within spheres, and the mechanical analogies of Ptolemy's epicycles and deferents that were used to calculate the future positions of the planets from knowledge of their past movements. Throughout the last 2,000 years, various mechanical models have been used to demonstrate the basic ideas of these mathematical and conceptual models.

The physical laws introduced by Sir Isaac Newton, combined with the powerful mathematical tools of differential and integral calculus, which he and Leibnitz invented, led to a cooling of interest in mechanical models. When scientists began to make discoveries about electric and magnetic fields, and how they were related to each other, many of them used mathematical techniques similar to those used by Newton in his treatment of the gravitation field. In his book *A History of Science* William Cecil Dampier said:

> The law of force being established, mathematicians took over the subject of electrostatics, and deduced an elaborate system of relations which proved concordant with observations wherever it was possible to make a comparison.... Of more historical importance was the attention directed to the electric force. Like gravitation, it appeared to act at a distance across intervening space. For mathematicians no further explanation was needed; but physicists soon began to speculate about the nature of this space, which somehow could transmit two, apparently distinct, forces. This led...to modern theories of what are now called "field physics."

As we saw in the Introduction and in Chapter 1, Michael Faraday, a man with very little mathematical knowledge, was in the forefront of this "field physics."

The Dutch theoretical physicist Hendrik Lorentz (1853–1928) also used models when he started to formulate how electromagnetic waves interacted with atoms of different elements. It was well known, at the end of the 19th century, that it was possible to identify a chemical element by the wavelengths of light it emitted or absorbed. At this stage in the history of science, no one knew what the atom looked like, but people knew of the existence of electrons, which were known to be much smaller than atoms. Lorentz assumed that the electrons within an atom behaved as damped simple oscillators. A mass swinging at the end of a string, or a small weight bobbing up and down at the end of a spiral spring, are two examples of simple harmonic oscillators. If the weight at the end of the spring was immersed in a liquid, then it would become a damped simple harmonic oscillator, because its movements would be damped by the liquid. The mathematics governing the behavior of damped simple harmonic oscillators is very similar, even though the physical situation may differ considerably. Lorentz assumed that within the atoms of a particular element there were electrons bobbing up and down on the ends of springs of different stiffnesses. Each different stiffness would cause an electron to emit or absorb light of a specific wavelength. The sets of stiffnesses would also vary from one element to the next. With such simple models Lorentz was able to provide a quantitative explanation for the behavior of atoms several years before the advent of the Bohr-Rutherford atom. He was also able to explain the Zeeman effect. Lorentz's theory was discussed by J.C. Slater and N.H. Frank in their book *Introduction to Theoretical Physics*: "In optics, the theory of refractive index and absorption coefficient is closely connected with resonance. As is shown by the sharp spectral lines, atoms contain oscillators capable of damped simple harmonic motion, or at any rate they act as if they did: the real theory, using wave mechanics, is complicated but leads essentially to this result."

The Rutherford-Bohr theory was initially based on an analogy with the solar system. Bohr's models made use of circular orbits, but the analogy with the solar system was taken further by Arnold Sommerfeld, who used elliptical orbits. Once the full quantum theory had been developed, there was much less reliance on physical models and analogies.

From about 1916 onwards, theoretical physics was to take a much more mathematical approach. This was largely due to Einstein's work on general relativity, and the work of Heisenberg and Dirac on quantum mechanics. Tensor calculus, group theory, symmetries, gauge theories, and supersymmetries took on a new importance. Now physicists began to dispense with pictorial and visual scaffolding, and instead concentrated on very abstract mathematical reasoning. I believe the present difficulties with string theory stem directly from ignoring other approaches that are more pictorial and visual.

The Milky Way Galaxy Model of Reality

The Milky Way galaxy is the most obvious feature of the night sky. All the stars we see belong to a vast disc-like distribution of stars, with a bulge near the center of the disc. The whitish, cloudy band of light that stretches across the sky comes from the light of many millions of stars, which are some of the more distant stars of the galaxy. The whitish gleam has several dark areas distributed among the bright parts of the band of light, due to dust clouds near the plane of the Milky Way.

Within the major part of the galaxy, the force of gravitation plays the dominant role. Gravitation holds the Milky Way together as a unit, but it also keeps all the stars orbiting around the center of the galaxy. The force of gravitation causes sufficient large clouds of gas and dust to collapse under their own mutual attraction to form clusters of stars. The most obvious feature of the Milky Way is the light from the stars, and these stars follow Newton's laws of gravitation and motion. The light from the stars interacts in two different ways with the dust particles, which form a relatively thin disc in the plane of the Galaxy (as we saw in Chapter 3). The dust particles not only reduce the amount of light we get from the stars, but they also polarize the light, and both effects are, to some extent, distant-dependant, although the polarization also depends on the direction in which we are looking. The dust is partly kept in the plane of the galaxy by gravitation, but the Galactic Magnetic Field controls the orientation of the dust grains, and it is this that causes the polarization of light. Thus the dust grains are under the control of two different types of forces: gravitation and magnetism.

As we also saw in Chapter 3, one of the early triumphs of radio astronomy was the detection of radio waves from the galaxy. These waves are generated by high-energy electrons spiraling around the lines of force of the Milky Way galaxy, and their speeds are so high that they are not affected much by the gravitational field of the galaxy, so it is the magnetic field of the galaxy that constrains them to move near the plane of the galaxy. However, the magnetic field itself is anchored in the gas and dust clouds of the Milky Way, so gravitation does, indirectly, keep the high-energy electrons near the plane. Although the electrons obey Newton's laws of motion, they are interacting with the magnetic field according to the laws of electrodynamics, which we discussed in Chapter 1. The point I wish to make is that the stars, the dust particles, and the charged particles of the Milky Way all follow slightly different sets of laws. This is not dissimilar from the points I made in the last chapter, where I posited that on a large scale, large particles and large bodies obeyed Newton's laws of motion, but on the subatomic scale, particles were affected by the plasma space world-line web of reality, which gave rise to different sets of laws for these particles. The presence of this world line web also generated additional attraction and mass (via the energy within it) on the galactic and galactic cluster scales, but on the cosmological scale the tension in the web was so much increased that the tension, or negative pressure, within sheaths of elasticated insulating space cause the expansion of the universe to accelerate according to the general theory of relativity.

The Urban Model of Reality

In 1984 Richard Trench and Ellis Hillman wrote a fascinating book called *London under London*, and in their preface they said, "Every time we turn on the tap, pull the chain, pick up the telephone, there is an underground movement: a gurgle of water, an impulse along a wire. As we bask in the electric sunshine of our city surface, we are quite unaware of the subterranean labyrinth honeycombing the ground beneath our feet. Very occasionally, in time of war, strike, or flood warning, we become aware of this troglodyte city, London under London. But **only** in exceptional times. In our everyday lives our ignorance of the world below extends to profound depths."

Although the book is specifically about subterranean London, some of the systems they discuss, such as water pipes and electric cables, are similar to those found in many urban environments. Others, such as the underground railway, are similar to those found in many large cities, for example Paris or New York.

I will start by listing some of the systems under London:

- Water and sewage systems.
- Gas mains.
- Electric, telecommunications, and cable television cables.
- Underground railways.
- Tunnels for pedestrians and cars.

Let us look at each of these systems in a bit more detail.

Water and Sewage Systems

A great deal of water, with varying degrees of cleanliness, flows under London. There are about12 rivers under this city, with a total length of about 100 miles. In their book, Trench and Hillman describe the process by which these rivers came to be buried: "The burying of London's rivers was a gradual process, beginning with the lower parts in the sixteenth century, when London burst out of its walls, spreading out towards the surrounding villages. The rivers formed natural valleys; valleys were natural roads; houses sprang up along roads; roads required drains, drains flowed into rivers; rivers became sewers; sewers became culverts."

The earliest water mains in London date from medieval times, and they were made of bored elm trunks, which were tapered at one end so that they could fit into one another. The elm trunks gave way to earthenware pipes, and these were eventually replaced by cast-iron pipes. The smaller pipes, leading from the mains in the street to the houses on either side, were made of lead.

Another system of water pipes was used under London from 1871 to the mid-1970s. These pipes were built to carry high-pressure water generated at five pumping stations around the city, and they were all owned by The London Hydraulic Power Company, which was founded in 1971. Trench and Hillman tell us that, "Raw (untreated) water was pumped at a pressure of 400 pounds per square inch through miles of pipes running beneath London, and was used to raise and lower cranes, operate lifts, West End theatre safety curtains, wagon hoists, even hat-blocking presses."

The use of this water power reached its heyday in 1927 when there were 184 miles of high-pressure pipes under London. The development of electric power started a slow decline in the demand for this type of power, but it was mainly the Second World War that caused a rapid decline in the fortunes of the company. During the London Blitz, 261 mains were broken. By the mid-1970s, the company ceased to operate.

Gas Mains

According to Trench and Hillman, "Gas was almost entirely a nineteenth-century phenomenon. The first street lighting by gas started in the 1800s, and the decline of the industry was underway by the 1890s."

Initially the gas companies followed the water companies in that they used bored-out elm trunks for their mains. This later gave way to earthenware and then to cast iron. However, they did not use lead pipes to take the gas from the mains to the meters, as the water companies did, but instead used army-surplus musket barrels, which were in cheap supply after the Napoleonic Wars. Although this practice has changed, these off-shoots from the mains are still referred to as "barrels" in the gas industry.

Electric, Telecommunications, and Cable Television Cables

The very rapid development of electricity for use in factories, office buildings, other commercial premises, and homes brought about the need for lines of transmission. The invention of the telephone added to this another set of lines. Trench and Hillman tell us that, "This, the nervous system of London, started life in the mid-nineteenth century, above ground. But overhead clutter, storm damage, and a train of accidents drove it underground about the turn of the century. It has been going deeper ever since."

Apparently, to begin with, the wires and cables were in ducts, which were rectangular and made of earthenware, about one yard below the level of the pavement, or sidewalk. The movement of traffic at ground level caused ripples in the ground below, and these ripples caused stretching of the wires and cables, and some of them eventually broke. This method also interfered with new building schemes, so it was decided to have them further below the surface. Both the London Electricity Board and British Telecom made their own deeper-level tunnels. Some of the cables and wires shared space with London Transport cables in the tunnels of the underground railway.

Underground Railways

New York, Barcelona, Paris, Sao Paulo, and Tokyo are among the great cities of the world that have underground systems. However, the world's first underground railway system was in London. At first, many people thought that running steam trains in tunnels under the streets of London was a stupid idea. However, the London underground turned out to be one of the great engineering achievements of modern times, the world's only steam-driven underground railway, and the world's first electrified underground railway.

Tunnels for Pedestrians and Motor Vehicles

There are also a few tunnels under the Thames, some of which are for pedestrians and some for motor vehicles. These were built at various times in the history of London, and they offer an alternative means of getting from one side of the river to the other, in addition to the bridges at ground level.

As with all towns and cities, there are a few systems above or at ground level, which are necessary for normal transport and communications. These include roads, streets, sidewalks, and radio links. The movement of traffic is prevented from being chaotic by having traffic laws, speed limits, traffic lights, and pedestrian crossings to enable people to cross busy roads, when there is no subway. Radio links are needed by the police, the fire service, ambulances, buses, and taxis.

The speed limits and the rules that govern the behavior of the different systems, do, obviously, differ from one system to the next. The speed limit for underground trains is different from that of cars above the ground, and the speed of the radio links is limited by the speed of light.

The main point I wish to make in this chapter is that in certain physical systems, such as our Milky Way galaxy, and in our towns and cities, it would be impossible to represent the behavior of the various subsystems by a unified mathematical model. However, we do not have to go into multidimensional spaces in order to understand our galaxy or to live and operate in our urban environments. Perhaps the search for a unified mathematical theory of space and time is just too much to ask. By making use of subspaces, within our normal framework of space and time, we can gain a workable understanding of the cosmos. The theory I proposed in Chapter 7 does imply that certain elements of the past are fossilized in the subspace, just as elements of the geological and biological history of our Earth are fossilized in the ground beneath our feet and the layers below the seabed. Some of the consequences of this fossilized past were explored in my book *The Third Level of Reality.*

Bibliography

Airy, George Biddell. "Tides and Waves." *Encyclopedia Metropolitan*. London: unknown publisher, 1845. Also appears in Lamb, Horace. *Hydrodynamics*. Cambridge, UK: Cambridge University Press, 1932, 254–270.

Bell, J.S. *Speakable and Unspeakable in Quantum Mechanics*. Cambridge, England: Cambridge University Press, 1987.

Berkson, William. *Fields of Force*. London: Routledge and Kegan Paul, 1974.

Bizony, Piers. *Atom*. Cambridge, England: Icon Books, 2007.

Blizard, J.B. "Long-Range Solar Flare Prediction: Final Report." NASA Contractor Report Number 61313, June 1968 to August 1969.

Bohm, David. *Wholeness and the Implicate Order*. London: Routledge and Kegan Paul, 1980.

Caldwell, Robert. "Dark Energy." *Physics World*, May 2004. *physicsworld.com/cws/article/print/19419* (accessed May 2008).

Cantor, Geoffrey, David Gooding, and Frank M.J.L. James. *Faraday*. London: MacMillan Education Ltd., 1991.

Clowe, Douglas, and Dennis Zaritsky. "Shot in the Dark." *Physics World* February 2007.

Dampier, William Cecil. *A History of Science*. Cambridge, England: Cambridge University Press, 1942.

Dar, Arno, and A. de Rujula. "Magnetic Fields in Galaxies, Galaxy Clusters, and the Intergalactic Space." *Physics. Rev.* D, 72, 123002, 2005.

Davies, Paul C.W. *The Cosmic Blueprint*. London: Unwin, 1989.

———. *Superstrings: A Theory of Everything*. Cambridge, England: Cambridge University Press, 1988.

Dyson, Freeman. *Disturbing the Universe*. New York: Harper and Row, 1979.

Eddington, Sir Arthur. *The Internal Constitution of Stars*. Cambridge, England: Cambridge University Press, 1988.

Einstein, Albert. "Maxwell's Influence on the Development of the Concept of Physical Reality." Written 1931. Taken from Mountain Man Graphics, *www.mountainman.com.au/aether_2.html*, fall 1997 (accessed May 2008).

———. *The Meaning of Relativity*. London: Routledge, 2003.

Eisberg, Robert Martin. *Fundamentals of Modern Physics*. New York: John Wiley and Sons, 1961.

Freire, G.F. "Force-Free Magnetic Configurations." *American Journal of Physics* 34 (1966): 567–570.

Gilbert, William. *De Magnete*. Mineola, New York: Dover Publications, 1991.

Gough, Douglas. "Letter to the Editor," *Nature* (1990), 345, 768.

Hamilton, James. *Faraday: The Life*. New York: Harper Collins, 2002.

Harman, P.M. *The Natural Philosophy of James Clerk Maxwell*. Cambridge, England: Cambridge University Press, 1998.

Hathaway, David. "Solar Cycle Predictions." NASA Solar Physics, Marshall Spaceflight Center, *solarscience.msfc.nasa.gov* (accessed May 2008).

Hoffman, Banesh. *Einstein.* St. Albans, Herts, England: Granata Publishing Ltd., Paladin Imprint, 1975.

Jones, C. "Making Sense of a Turbulent Universe." *Physics World* 8, 7 (1995): 50–51.

Jose, P.D. "Sun's Motion and Sunspots." *The Astronomical Journal* 70 (1965): 193–200.

Kargon, Robert. *Science in Victorian Manchester.* Manchester, England: Manchester University Press, 1977.

Manchester, R.N., J.L. Han, A.G. Lyne, G.J. Qiao, and W. van Straten. "Pulsar Rotation Measures and the Large-Scale Structure of the Galactic Magnetic Field." *Astrophysical Journal* 642 (May 10, 2006): 868–881.

Maxwell, James Clerk. *A Treatise on Electricity and Magnetism.* Oxford, England: Oxford Classical Texts in the Physical Sciences, 1998.

Mercet, Jane. *Conversations on Chemistry.* Hartford, Conn.: Unknown publisher, 1809.

Moffat, K. "Cosmic Dynamos: From Alpha to Omega." *Physics World* 6, 5 (1993): 38–42.

Nicholson, Iain. *The Dark Side of the Universe.* Bristol, England: Canopus Publishing Limited, 2007.

Parker, E.N. *Cosmical Magnetic Fields: Their Origin and Their Activity.* Oxford, England: Clarendon Press, 1979.

Polya, G. *How to Solve It.* Princeton, N.J.: Princeton University Press, 1998.

Powell, David. "Moon's Magnetic Umbrella Seen as Safe Haven for Explorers." Science Tuesday, *www.space.com/scienceastronomy/061114_reiner_gamma.html*, posted 11/14/06 (accessed May 2008).

Sanders, Robert. "Strong Evidence that Mars Once Had an Ocean." University of California, Berkeley, Press Release, June 13, 2007.

Seymour, Percy. *Cosmic Magnetism.* Boston: Adam Hilger, 1986.

———. "The Elementary Particles as Stable and Unstable Localized Energy Modes in Electrified Space-Time." *Occasional Papers in Astrophysics* 1, No. 2, Plymouth Polytechnic, 1989.

———. *The Scientific Basis of Astrology*. New York: St. Martin's Press, 1992.

———. *The Third Level of Reality*. New York: Paraview, 2003.

Seymour, Percy, and M.J. Willmott. "A New Theory for the Solar Cycle." *Journal of the National Council for Geocosmic Research* Winter 2000–2001.

Seymour, Percy, M.J. Willmott, and A. Turner. "Sunspots, Planetary Alignments, and Solar Magnetism." *Vistas in Astronomy* 35 (1992): 39–71.

Sherwood, Jonathan. "Lecturer describes what may have happened just before the big bang." *Cornell Chronicle* 29, No. 25, March 12, 1998.

Slater, J.C., and N.H. Frank. *Introduction to Theoretical Physics.* New York: McGraw-Hill, 1933.

Smolin, Lee. *The Trouble With Physics: The Rise of String Theory, the Fall of Science and What Comes Next.* New York: Penguin Books, 2007.

Trench, Richard, and Ellis Hillman. *London under London: A Subterranean Guide.* London: John Murray, 1984.

Whitehead, Alfred North. *Science and the Modern World.* Cambridge, England: Cambridge University Press, 1929.

Wilmot-Smith, A. L., P.C.H. Martens, D. Nandy, E.R. Priest, and S.M. Tobias. "Low-Order Stellar Dynamo Models." *Monthly Notices of the Royal Astronomical Society*, 363, 4, (2005):1167-1172.

Index

About the Author

Percy Seymour was born in Kimberley, South Africa, where, nurtured by his grandfather, he took a keen interest in astronomy at an early age, under the clear skies of the African veldt. He received his school education in Kimberley, and spent some time at the University of the Witwatersrand in Johannesburg, but completed his studies at Manchester University in England, where he specialized in physics and astrophysics. In his research for higher degrees, he pioneered the use of mathematical techniques, first used to study the magnetic field of Earth, to investigate the magnetic field of the Milky Way galaxy.

After a few years as a schoolteacher, he became senior planetarium lecturer at the Royal Observatory, Greenwich, the home of the Prime Meridian of the World. Before his retirement in 2003 he was principal lecturer in astronomy at the University of Plymouth in the UK. Besides several scientific papers, mainly on magnetic fields in astronomy, he has also written numerous articles for magazines and newspapers. In 1984 he made world news when his work on the Star of Bethlehem at the planetarium in Plymouth was reported on the front page of *The Times*. His books include *The Birth of Christ, Halley's Comet, Adventures with Astronomy, Cosmic Magnetism, Astrology: The Evidence of Science*, and *The Third Level of Reality*. Some of these books have been translated into German, Dutch, Italian, Portuguese, and Arabic. He is married, with one son, and currently lives in Somerset, England.